PHIL MELLOWS
AND
KATE SIMON

BEER BREAKS IN BRITAIN

30 PLACES TO EXPLORE AND DRINK GOOD BEER

CONWAY

LONDON · OXFORD · NEW YORK · NEW DELHI · SYDNEY

CONWAY
Bloomsbury Publishing Plc
50 Bedford Square, London, WC1B 3DP, UK
29 Earlsfort Terrace, Dublin 2 , Ireland

BLOOMSBURY, CONWAY and the Conway logo are trademarks of Bloomsbury Publishing Plc

First published in Great Britain 2025

A catalogue record for this book is available from the British Library

ISBN: TPB: 978-1-8448-6700-4; eBook: 978-1-8448-6701-1 ePDF: 978-1-8448-6698-4

2 4 6 8 10 9 7 5 3 1

Designed by Paul Palmer-Edwards

Printed and bound in India by Replika Press Pvt. Ltd.

To find out more about our authors and books
visit www.bloomsbury.com and sign up for our newsletters

CONTENTS

BREWED BY US
EMPEROR PALE ALE 4.4% £4.85
GLADIATOR BEST BITTER 3.8% £4.60
LIONESS HAZY IPA 4.9% £5
CAPRICE PILSNER 5.6% £5.70
SAKURA RICE LAGER 5.6% £5.60
NORSE KVEIK PALE ALE 5% £6.10
OPTIC NEW ENGLAND SESSION IPA 3.6% £5.20
WARRIOR INDIA PALE ALE 5.5% £5.10
CARPE NOCTEM DARK GERMAN BEER 5.5% £4.85
GUEST BEER
LOST & GROUNDED KELLER PILS 4.8% HOP BITTER £5.40 LAGER
NORTHERN MONK eternal SESSION IPA 4.1% 330ml can
JUBEL Session Lager WITH PEACH 4% £4.90 330ml Can
BULLHOUSE BREW CO. HAZY IPA ROLLING PAPERS £6.20 5.2%
NORTHERN MONK "faith" HAZY PALE IPA 5.4% 440ml can
tasty juice EAST COAST IPA 6% 500ml can
COMMODUS
CASSIUS
CAESAR
SUFFRAJITSU
HERCULES
MINOTAUR
DEUS
LUPULUS
EMPRESS
AUGUSTUS
HADRIAN
THE BATH BREW HOUSE
GLADIATOR BITTER
THE BATH BREW HOUSE
EMPEROR PALE ALE 4.4%
THE BATH BREW HOUSE
WARRIOR INDIA PALE ALE 5.5%
THE BATH BREW HOUSE
ANGRAM
rhino
STU DIS
LID I.D. REQUI

ACKNOWLEDGEMENTS

This book is dedicated to Simone Kane, our former partner in all things beery, and Campbell Reid, whose timely thought sparked the initial idea.

Visiting 30-plus destinations and hundreds of pubs, bars and breweries is no small undertaking and we are very grateful for the hosting and the extra assistance we received from Avanti West Coast (www.avantiwestcoast.co.uk), LNER (www.lner.co.uk), Rose Davis and Shepherd Neame (www.shepherdneame.co.uk), Andy Milburn and St Austell Brewery (https://staustellbrewery.co.uk), Inn Collection Group (www.inncollectiongroup.com), Bespoke Inns (https://bespokeinns.co.uk/), Old Hall Inn (www.old-hall-inn.co.uk), Louise Ferrall at Visit Britain (www.visitbritain.com), Samantha Marsh and Cumbria Tourism (www.visitlakedistrict.com), Simon Blenkinsop at Cumbria Tourist Guides (www.cumbriatouristguides.org), Mark Hibbert, Lawrence Oates and Discover East Staffordshire (https://discovereaststaffordshire.com), Michelle Booth and Visit Derby (www.visitderby.co.uk), Amy Noton and Visit Peak District & Derbyshire (https://visitpeakdistrict.com), Visit Cornwall (www.visitcornwall.com), Katie Bentley at Liverpool BID, West Midlands Growth Company (https://wmgrowth.com), Glasgow Life (www.glasgowlife.org.uk), Jackie Ellis at Tourism Angles (www.tourism-angles.co.uk), Sophie Barber at Kelham Island Food Tours (www.kelhamislandfoodtours.co.uk), Donnington Brewery (https://donnington-brewery.com), and Fergus Fitzgerald and Adnams Brewery (https://adnams.co.uk).

Kate would also like to thank her mum, Sted Simon, Dean Ryan, Gordon Jamieson, Joanna Fernández, Rachel Graham, Isabel Lewis, Ken Mulkearn, Lottie Gross, Sarah Baxter, Jo Barrell, Suzanne King, Lou and Gareth Gaskell, Michele Holmes, Marcus Field, Danny Budzak at https://commodityfetishism.com and everyone else who offered suggestions worth investigating.

Phil would also like to thank Paul Mellows, Nick Russen, Richard Brunsden, Pete and Jackie Martin-Green, Paul White and Alison Howard, Jezz Etheridge and Sallie Richards, Kate Hale, Mark Dorber, Steve Hobman, Pam Lock, Emma Inch, Matt Curtis, Eddie Gadd, Duncan Sambrook and, most of all, Google Maps. Over the years many people have recommended pubs to Phil, and he thanks them, too. Cheers!

Opposite: Bath, Brew House

FOREWORD

As a beer sommelier with a penchant for pints and a love for all things hops and barley, Beer Breaks in Britain is my kind of travel guide!

Phil and Kate have crafted the ultimate pub crawl without the hangover – unless, of course, you follow their advice too well. From hidden gems to iconic brews, this book is your passport to the best beer spots in Britain.

Whether you're a stout supporter, a lager lover, or just here for the ales, this guide will have you saying 'cheers!' in no time.

Remember, life's too short to drink bad beer – or to miss out on this fantastic read!

Marverine Cole, award-winning journalist, broadcaster and beer sommelier

SCOTLAND
Glasgow
Edinburgh
Lake District
THE NORTH
NORTH WEST
Keighley to Shipley
York
Leeds
Manchester
Sheffield
Liverpool
Chester
Peak District
North Wales Coast
Derby
Nottingham
Burton Upon Trent
MIDLANDS
WALES
Birmingham
Norwich
Southwold & Walberswick
LONDON & EAST
Pembrokeshire
Cheltenham
The Cotswolds
Hackney
Walthamstow
Isle of Thanet
Bristol
Bermondsey
Bath
Faversham
Canterbury
SOUTH
SOUTH WEST
Brighton
Lewes
Cornwall

INTRODUCTION

Beer takes you places. The search for good beer can get you under the skin of a city or into the remote corners of the countryside. Seeking out the wide variety of styles now available can land you in historic pubs and modern bars, brewery taprooms and somebody else's local.

In this book we join the dots between beer and travel, a perfect partnership we have been writing about for our *British Beer Breaks* newsletter on Substack since 2022 (http://britishbeerbreaks. substack.com). We've focused on 30 destinations in England, Wales and Scotland where you can drink good beer and, when you're not, find interesting things to see and do to complete a satisfying break.

Our aim is to inspire. You won't find an exhaustive list of beer and tourism venues and experiences; instead, we celebrate locality and sustainability, selecting what we've enjoyed and think is worth sharing with you. We don't have a narrow definition of what makes good beer. For a long time in the UK it rightly meant cask ale. Since the craft revolution that rolled on to these shores around 2010, it has broadened immensely, drawing in a more diverse audience, keen to experiment and explore.

Each of the beer venues covered in this book has something special, whether it's the wide range of beer, or the opportunity to try a beer you may not easily find elsewhere or that is closely associated with the destination. Sometimes you can even learn something new about a place from the name of a local beer. You will, no doubt, discover more on your travels, but we hope we can help cut down on the trial and error usually required to find good beer in an unfamiliar place.

Our choices follow a geography that suggests how you can move between them, but it's up to you to curate and pace your own beer break, managing your drinking by choosing lower-strength beers, including alcohol-free options that are improving all the time, smaller measures such as two-thirds and thirds of a pint, as well as halves, and remembering to stop to eat.

The idea behind this book is that while you're having a few beers you can also enjoy the other things a destination has to offer. Our whole approach is underpinned by the principle of Slow Travel. This means taking time to appreciate and immerse yourself in the place you're visiting by meeting the people, sampling the area's food and, of course, the drink, discovering local arts and crafts, learning about history, culture and nature, enjoying the built and natural landscapes and, where possible, exploring on foot, bicycle and by public transport.

We've travelled to all these destinations, but some we know better than others. Where we aren't as familiar with a place, we've gathered recommendations from trusted friends and contacts. Our aim has been to seek out the lesser-known things to do and see as well as the obvious yet unmissable. We can't cover everything – this is our pick – but you'll find a link to the local tourist board for further information.

There are also some pointers for beer-themed places in which to stay, including several reviews of pubs with rooms. Take it as read that where rooms are located above the bar, noise is likely to travel up from below.

We hope you enjoy our book and that it will prove an indispensable resource for marrying beer and travel to create rewarding short breaks.

Opposite: Beer Shed, New Mills.

Beer Shed
Opening Times
MON ~CLOSED~
TUE ~CLOSED~
WED 16:00~22:30
THU 16:00~22:30
FRI 14:00~22:30
SAT 14:00~22:30
SUN 14:00~21:30
CASK CRAFT BEER
RUM · GIN
WHISKY
FINE WINES
HAVANA CIGARS
local cask ale
micro pub

CASK & CRAFT

Cask ale is a 'living' beer that completes its manufacture in the cellar of the pub. It's found in very few places outside Britain. The quality of the beer in the glass is as much down to the skill and timing of the publican as the brewer. And it has a limited shelf-life once tapped. It's most easily recognised by the way it's poured – typically from long handpumps on the bar, or occasionally straight from the barrel. The perfect pint should be cool and spritzy, certainly not warm and flat.

The prevalent method of draught dispense around the world is from a pressurised keg. Poured from a simple tap, it requires less intervention from the publican and has a much longer shelf-life. Mostly, it's used for lager, but recent years have brought an explosion of high-quality 'craft' beers from small independent producers in all manner of styles. A couple of leading brands have become ubiquitous, and less special, so we've focused our attention on the more interesting brews that you won't find in almost any local.

It's worth noting that there is no longer a sharp divide between cask and craft. Traditional family brewers that have been around for centuries are now producing their own takes on 'craft', while cutting-edge modern brewers increasingly make cask ale. And there is a growing culture of collaboration between old and new.

NATIONAL CHAINS

We have deliberately skipped national pub and bar chains to better highlight local venues. The main ones are **J D Wetherspoon** (www.jdwetherspoon.com), which typically occupies prominent, and often architecturally interesting, buildings in many towns and cities, and usually serves a decent range of cask ale, and **Brewdog** (www.brewdog.com), which, from small beginnings, is now a global company with 70 bars in the UK pouring craft beers from a variety of independent brewers alongside its own popular selections.

Opposite: Innis & Gunn taproom, Edinburgh.

STAFF FAVOURITE
5.6%   PINT   £6.4
5.0%   PINT   £6.4
ZEN   5.5%   1/2 PINT   £6.3
TEN FREE VEGAN LAGER   4.0%   PINT   £7.5
K MATURED CHERRY BEER   5.1%   2/3 PINT   £5.2
LE CIDER   4.5%   PINT   £5.9
DEN ALE   4.1%   PINT   £5.7
TMEAL STOUT   4.1%   PINT   £5.7
ER   4.6%   PINT   £5.9
38%   1/2 PINT   £40.8
8.4%   1/3   £5.2
SERT SOUR   LOCAL TAP   6.5%   1/3   £5.2
NACHAN PALE ALE   5.5%   1/2 PINT   £6.
GO & PASSION FRUIT   10.0%   1/3   £9.9
SSION SOUR   4.1%   1/2 PINT   £5.5
Y NEIPA   6.0%   1/2 PINT   £5.2
TEN FREE PILSNER   P   4.8%   2/3 PINT   £4.9
TEN FREE NITRO   4.5%   PINT   £7.0
STRY SOUR   6.8%   1/3   £5.0
PBERRY SOUR   FA   4.0%   2/3 PINT   £5.3
EA E   5.0%   2/3 PINT   £5.3

THE NORTH

Opposite: River Don, Sheffield.

YORK

York's pubs and bars exhibit all the historic, architectural interest you'd expect in a city like this – plus a dash of cutting-edge craft thanks to Brew York. So do its attractions, from the Minster to the Shambles, but there are a few more modern surprises on the tourist trail, too. And it can all be experienced in one conveniently compact circuit.

 WHERE TO DRINK

Inside the railway station, **York Tap** is the perfect place to begin, or end, your tour. It occupies a spectacular Edwardian building with stained glass windows and Art Nouveau detail you can gaze upon while drinking your choice of beer from up to 20 cask pumps pouring traditional and modern styles, plus 10 keg taps that combine British craft with continental lagers and wheat beer. https://yorktap.com.

Below the city walls by Lendal Bridge, **The Maltings** is a characterful pub with a rustic feel and walls adorned with retro advertising plates. Eight cask lines lead on Black Sheep ales and there are half a dozen craft selections, too. www.maltings.co.uk.

The first **House of the Trembling Madness** you'll encounter in York is across the bridge. It describes itself as a 'craft beer mansion' with good reason, housed as it is over three floors of a splendid Georgian building previously occupied by a long list of Yorkshire gentry. The bar at the front doubles as a café and offers 11 keg lines featuring British and continental brewers, plus three more on cask. The original **Madness** among the quaint shops of Stonegate is, while smaller, even more impressive. The bar is above the company's ground-floor bottle shop in a building that dates back to the 12th century and oozes history from its beams. The beers pouring from its eight keg lines and three cask, though, are thoroughly modern. www.tremblingmadness.co.uk.

Over in the Shambles, **Pivni** was a pioneer of the UK's craft beer movement, opening in this three-storey 16th-century building as early as 2007. Today, it's still bringing together cask and the best of keg on a total of 15 lines, the beers stored in an unfeasibly cramped cold room to the side of the bar. **Ye Old Shambles Tavern** is an odd little place, on one side opening onto a beer terrace in the old Shambles Market, and on the other a cute little bar that's the best place to sample the house ales brewed by York's Rudgate and a few other good choices. On the corner is Thornbridge's **Market Cat**, a recent refurbishment over three storeys with a great view of the city from the top. Eight cask lines feature brewers perhaps better known for their craft beers while 14 keg taps complete a generous menu. https://pivni.co.uk. www.shamblestavern.com. www.marketcatyork.co.uk.

[1]

The tiny **Blue Bell** is a York legend, dating from 1798 and proudly declaring it hasn't been redecorated since 1903. Two rooms and a corridor surround a central bar serving six cask ales. Opposite, Ossett Brewery has converted a rather grand Italian restaurant into **The Hop**, combining good beer, pizzas and live entertainment. Six Ossett cask ales line up alongside craft beers from its Salt Brewing arm. https://bluebellyork.com. www.hopyork.co.uk.

Brew York Beer Hall & Tap Room [1] occupies a riverside warehouse off Walmgate, sharing the site with the firm's original brewery. The tap room is downstairs and opens onto a terrace with views of the Foss. Six handpumps give a rare opportunity to try Brew York's beers on cask and there are 10 keg lines, too. But that's puny compared to the beer hall upstairs, where you can choose from 40 craft beers on tap, including eight or 10 guests, all tidily listed by style on a screen above the bar. Brewery tours and tastings are available. https://brewyork.co.uk/venues/walmgate. www.sparkyork.org.

Out by the city walls to the south, three pubs are worth visiting. The **Rook & Gaskill** has a micropub feel to it, though it's too large for that. Up to eight cask ales are on the bar, including Bateman's XB, plus a good choice of craft on draught and more in the fridge. The **Phoenix Inn** is a quiet, uncomplicated freehouse serving five sound beers on cask, a proper local pub. As is the **Swan**, a real beauty with its multiroomed interwar interior where you can enjoy your choice of eight handpulled cask ales. www.rookandgaskillyork.co.uk. www.phoenixinnyork.co.uk. www.theswanyork.co.uk.

The **Golden Ball** is a Grade II-listed pub that dates back to 1773 and was remodelled between the wars. Some unique features have survived, including the bar-side alcove seating. Most importantly, the business was taken over by the local community in 2012 and has become a vital local hub. The beer's good, too, with half a dozen cask ales pouring. www.goldenballyork.co.uk.

[2]

👓 WHAT TO SEE & DO

They like to dress up in York. But when your tourist attraction has so many others to compete with you've got to stand apart somehow. At the **Jorvik Viking Centre**, where the clock stopped in 975 AD, they wear capes and carry shields, and anything from a tricorn to a boater is worn with appropriate clobber from down the centuries at **York Castle Museum** [2].
www.jorvikvikingcentre.co.uk.
www.yorkcastlemuseum.org.uk.

The sights just keep coming from the moment you step out of the railway station. Cross the road to join the **city walls**, the most complete medieval circuit in Britain. They lasso a huddle of buildings that record human presence here at the confluence of the rivers **Ouse** and **Foss** since Roman times. Barely a few hundred metres ahead, drop down by the **Skeldergate Bridge**, designed by the architect of London's Westminster Bridge, then descend the stairs by the once-crucial medieval defence **Lendal Tower** (which you can stay in) to the banks of the Ouse, continuing

forward along Dame Judi Dench Walk (she was born here) to **Museum Gardens**.

Here, not only will you find Victorian botanical gardens planted by the Yorkshire Philosophical Society but also the early museum of its type, which they built to house their archaeological and natural treasures, now including the **Ryedale Roman Hoard**, four fragile miniature objects featuring a haunting hollow-eyed bust of the emperor Marcus Aurelius that once capped a sceptre. There is also a medieval hospital, a groundbreaking Victorian observatory, and a corner of the Roman fortress in these grounds, although the most arresting vision is the drift of ruins of **St Mary's Abbey Church**.
https://lendaltower.com.
www.yorkshiremuseum.org.uk.
www.yorkmuseumgardens.org.uk.

Take the path into **Exhibition Square** to see the **King's Manor**, a nerve centre of power in the north for centuries. It's part of the university so rarely open to the public, but you can admire Charles I's florid coat of arms over the entrance. Next door, **York Art Gallery** regularly reappreciates some of its permanent works from different perspectives. On last visit, the curators had arranged an interesting LGBTQ+ take on paintings from Juan Carreño de Miranda's mid-17th century depiction of Saint Sebastian to Grayson Perry's 2014 body-positive ceramic bust, *Melanie*. www.yorkartgallery.org.uk.

Pass through the city gate **Bootham Bar** and you're into the tangle of streets, which, like Bath, has UNESCO status, to find the medieval alleys with names such as Nether Hornpot Lane known as the **Snickelways** – a folksy term in fact coined in the 1980s by a local author. This is where you'll find the famous **Shambles**, a tight cobbled street with overhanging buildings where Harry Potter fans (the Shambles is believed to be the inspiration for the movie adaptation's Diagon Alley) and Instagram wannabes hang out. Buck the trend and check out the **typographical mural** behind Browns department store. Not a sight you expect to see. It's by the Los Angeles-based street artists Defer and Big Sleeps, created in collaboration with York's **Art of Protest Gallery**. www.artofprotestgallery.com.

Back on the tourist trail, it's time to tackle the main attractions – there's a mound to climb to Henry III's **Clifford's Tower**, the medieval keep of the castle, as well as the timber-framed **Merchant Adventurers' Hall** to see, from which an influential guild operated, and the mighty Gothic **York Minster**. www.english-heritage.org.uk. www.merchantshallyork.org. https://yorkminster.org.

WHILE YOU'RE HERE

The 18th-century **Holgate Windmill** is said to be the finest five-sail specimen in Britain, while your inner child will thank you for visiting the **National Railway Museum**. www.holgatewindmill.org. www.railwaymuseum.org.uk.

FURTHER INFORMATION

For more ideas on what to see and do, go to https://visityork.org. Check opening times at beer venues and attractions before you go.

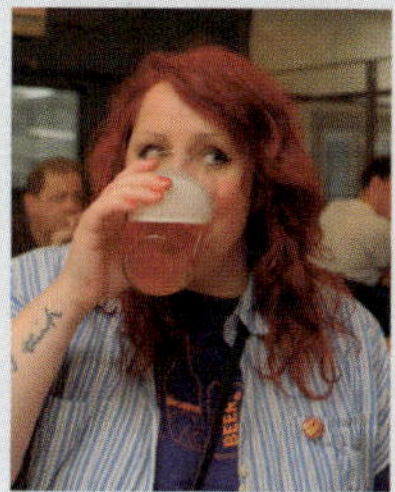

'Women belong in beer' runs the slogan of Women on Tap, and that's certainly where Rachel Auty, the organisation's founder and former head of marketing at Brew York, feels at home.

'I got into beer as a student when it was very much brown compared with the different styles you get now. But I still loved it back then. I like the social thing around it, the pint in the local pub. I've always enjoyed that.'

Seeing the need to overcome the stigma attached to women drinking beer, Rachel launched Women on Tap in 2017, which organises festivals and raises awareness of the issue.

'I started talking to brewers and out of that came some consultancy work for Brew York, which then offered me a job just before the pandemic. As well as liaising with the media I helped organise the company's Brewery Bash each April. It takes over the whole site in York and 2,000 people come during the two days.'

Although she lives in Harrogate, Rachel will often hop on the train for a pint at the York Tap, and she believes the city has a lot to offer when it comes to beer.

'It's a vibrant scene,' she says. 'Brew York's Tap Room and Beer Hall are a honeypot for visitors and there are innovative places like the House of Trembling Madness, quirky bars and traditional pubs such as the Maltings, which still has some crazy pours. There's so much choice, and you're surrounded by history, too.'

www.womenontap.co.uk.

LEEDS

As well as being home to the historic Whitelock's Ale House, Leeds was a forerunner of the modern beer movement with the opening of North Bar in 1997 and is now a major centre of craft brewing. It's also a major centre of commerce, as revealed by its network of elegant Victorian shopping arcades.

 WHERE TO DRINK

Aptly, the first place you hit walking from the railway station is Leeds Brewery's **Brewery Tap**. No beers are brewed on site here, but there's a wide range of craft and cask, including plenty of guests from around Yorkshire. Enjoy them in a relaxed, comfortable space with a roof terrace above. https://brewerytapleeds.co.uk.

The **Head of Steam** branch on Mill Hill [3] is one of the cosier examples of the brand, arranged around a circular island bar with a gallery above. From there you get a full sense of the wide range of beers stocked, both British and foreign, and a view of the unusual spiral chandelier of green bottles. Bang opposite is the original **Bundobust** bar-restaurant where the craft beer and Indian street food concept was born.
www.theheadofsteam.co.uk/bars/leeds-city-centre.
https://bundobust.com/locations/leeds.

A harmonious fusion of craft beer bar and Spanish café over two floors, **Friends of Ham** is a relaxed and welcoming place to while away an hour. A dozen lines of exceptionally well-chosen brews, from the likes of Vault City, Verdant, Cloudwater and alcohol-free specialist Mash Gang, join a couple on cask. Tutored tastings are available matching the beers with top quality cheeses and charcuterie.
https://friendsofham.com.

Thornbridge Brewery's joint venture with Pivovar, the **Bankers Cat** [4], is a spectacular place to drink beer. Housed in a grand former bank, it has a stained-glass window replicating the one by Burne-Jones in Thornbridge Hall, a Grade II-listed stately home in the Derbyshire Dales. Choose from eight cask lines and some interesting craft options as you sink into your leather sofa beneath a glittering chandelier. Pivovar is also behind **Tapped**, a modern brewpub with the tanks ranged down one side and the bar on the other. Geared to introducing the latest in beer to the people of Leeds, the home brews are supplemented with plenty of tempting options from further afield, across 13 keg lines and eight cask.
www.bankerscat.co.uk.
https://tappedleeds.co.uk.

Come out of the 'secret' entrance at the back of Leeds Station and you'll emerge at Granary Wharf, where Ossett Brewery appears to be making some kind of takeover bid. As well as

INNOVATIONE
PASSIONIS
COGNITIO

[5]

The Hop, a traditional-style pub with the firm's full range, and the neighbouring **Archie's Bar and Kitchen**, there's a waterside **Salt Taproom** from its craft beer arm, bristling with taps and a falafel pop-up. A short cruise downriver you'll find **Salt Calls Landing**, a more comfortable spot where you can enjoy a similar range of beers while taking in a fine view of the River Aire. While you're there, it's worth checking out **White Cloth Hall**, a food and drink hall opened in August 2024 by the people behind Whitelock's. https://thehopleeds.co.uk. www.facebook.com/ArchiesLeeds/?locale=en_GB. https://taps.saltbeerfactory.co.uk.

The **North Taproom** in Sovereign Street is a buzzing bar with a terrific range of craft keg on its 24 lines including the brewer's latest creations and intriguing collabs. On Temple, behind Granary Wharf, over two floors, the **Midnight Bell** is Leeds Brewery's flagship pub and a good place to sample its brews on cask alongside local guests. https://info.northbrewing.com/venue/sovereign-street. Midnight Bell, https://midnightbell.co.uk.

The city's second **Head of Steam**, among the shops and offices on Park Row, is designed on the new model for the chain, more open with a modern sheen. The beer list is just as extensive and inclusive and it has a separate gaming area. www.theheadofsteam.co.uk/bars/leeds-park-row.

Opening in 1715 as the Turk's Head, **Whitelock's Ale House** [5] has somehow survived the transformation of Leeds city centre into a vast shopping mall and can be found down an alleyway behind a cluster of jewellery stores. This gem of a pub breathes history and beer

[6]

culture with house cask ales including Kirkstall Pale, Theakston's Old Peculiar and a couple of Five Points brews all the way from Hackney, alongside four guests, mostly from Yorkshire. All that, plus there are five taps dedicated to craft keg, too. https://whitelocksleeds.com.

Salt x Hubb is an interesting collaboration between Salt Brewing and an artisanal coffee shop that's the brainchild of its former managing director. A coffee bar at the front opens in the morning and the venue evolves during the day into a night spot pouring seven craft beers and a couple of cask ales. Behind Leeds Art Gallery,

Northern Market [6] brings together street food vendors with a beer hall. Northern Monk supplies the liquids from 10 craft taps. https://taps.saltbeerfactory.co.uk/leeds-city. www.northernmonk.com/pages/great-northern-market.

The modest spot where North Brewing was born, **North Bar** was among the first in the UK to combine cask ale with Belgian classics and what were then the new-fangled American boutique beers. It's lost none of its vibrancy and passion for a variety of styles on both cask and keg. https://info.northbrewing.com/venue/north-bar.

🔭 WHAT TO SEE & DO

We won't take you to the shops very often in this book, but Leeds is an exception. The city has a long history of producing and exporting goods, and selling them in style, too. Like most cities, there is a hectic shopping area, but within it are several elegant **Victorian arcades**.

Thornton's Arcade was the first to open in 1878, the initiative of Charles Thornton, a pub landlord turned theatrical entrepreneur whose **City Varieties** is still open on Swan Street, now the longest running music hall in the UK. Thornton commissioned George Smith, who designed his theatre, to build the arcade (all the rage on the Continent at the time) to provide the good people of Leeds with a place out of the rain in which to part with their cash. And they still do.

The tall and slim passage is lined with handsome wood-framed shopfronts and topped with a span of glass supported by blue wrought-iron arches decorated with red dragons. You could set your watch by its Ivanhoe Clock, where life-size wooden figures from Sir Walter Scott's epic tale – Robin Hood, Friar Tuck, Richard the Lionheart and, from the supporting cast, Gurth the Swineherd – strike the bells.

Your shopping choices in Thornton's include a comic store, a tattooist and a tailor. The **County Arcade** [7] is rather smarter. Taller, wider and grander, too. It has an even bigger arched and domed glass roof; even more detail in its design, with balconies, bays and arcades; and even more ornamentation to peer at, from the wrought iron and painted plaster above your head to the mosaics of tiles beneath your feet. No wonder Louis Vuitton, Vivienne Westwood and the like have pitched up here. The other impressive thoroughfares to browse are **Grand Arcade**, **Queens Arcade** and the **Victoria Quarter**. www.victorialeeds.co.uk. www.queensarcade.co.uk. https://leedsheritagetheatres.com.

Kirkgate Market received similarly ostentatious treatment when it was remodelled at the turn of the 20th century. Among the hundreds of stalls is the original **Marks & Spencer Penny Bazaar**, a national retail institution that began trading here

in 1884. (The **M&S Archive** at the University of Leeds is a fond trip down memory lane, with St Michael's undies on display and a film theatre showing sexist ad campaigns from the Sixties and Seventies that will raise your eyebrows.) www.leeds.gov.uk/leedsmarkets. https://archive.marksandspencer.com.

The oval-shaped **Corn Exchange** is another remarkable shrine to commerce. Modelled on the French stock exchange, its pillared porches have welcomed traders since the 1860s – it's said the roof was designed so that the fall of the light ensured merchants could properly scrutinise the quality of the corn. Now independent retailers sell an eclectic selection of goods in the rainbow-coloured shops. The Corn Exchange sits close to **The Calls**, a cobbled road along the once-busy docks, the warehouses of which have been renovated into swish waterside apartments, offices, shops and places to stay, eat and drink. For more about Leeds' important role as a hothouse of engineering and manufacture, from steam locomotives to ready-to-wear clothes, head to **Leeds City Museum** and **Leeds Industrial Museum** at Armley Mills. www.leedscornexchange.co.uk. https://museumsandgalleries.leeds.gov.uk/ leeds-city-museum. https://museumsandgalleries.leeds.gov.uk/ leeds-industrial-museum.

TETLEY – BACK IN BEER

Tetley's Brewery has remained a prominent and much-loved landmark in Leeds, even though it ceased production in 2011. Until 2023 the Art Deco building did service as an art gallery – then **Kirkstall Brewery** announced it had taken on the lease and intended to restore its place at the heart of the city's beer culture. Now simply called **The Tetley**, it reopened as a pub in the spring of 2024, showcasing the best of Leeds brewing, including Kirkstall's own beers. Part of the Aire Park redevelopment, it is surrounded by an eight-acre park that includes Tetley Green and Theatre Gardens.
Aire Park, www.airpark.co.uk

Many of these latter-day expressions of civic pride in bricks and mortar are being crowded out by new buildings of questionable architectural worth. The maquette of one modern brick construction that perhaps should have made it off the drawing board in the 1980s but didn't due to local objections is on show at **Leeds Art Gallery**. Sir Antony Gormley's 37m (100ft)-high figure *Brick Man*, with an entrance in his heel and a viewing window in his head, would have stood by the railway tracks welcoming train passengers into the station – the burghers of Leeds must be kicking themselves all the way to the *Angel of the North*.

The gallery is building one of the leading collections of British sculpture in the world in collaboration with the neighbouring **Henry Moore Institute**, so there is always something fresh to see, as well as the room dedicated to works by Moore, fellow Yorkshire-born sculptor Barbara Hepworth and other Modernists. Sculptures by Moore and Hepworth are among 17 works from the 20th and 21st centuries displayed outdoors at the University of Leeds that have been linked together in the **Curating The Campus** art trail, a double delight because it weaves through the Victorian and Brutalist complex.
https://museumsandgalleries.leeds.gov.uk/
    leeds-art-gallery.
https://henry-moore.org/henry-moore-institute.
Curating The Campus: Public Art Trail, available
    from The Stanley & Audrey Burton Gallery,
https://library.leeds.ac.uk/info/1900/galleries.

FURTHER INFORMATION

For more ideas on what to see and do, go to www.visitleeds.co.uk. Check opening times at beer venues and attractions before you go.

Tiled Hall Café: Stop for a coffee at Leeds Central Library (where Alan Bennett used to do his homework) and, as you sip, admire the beautiful Victorian tiles on the walls and vaulted ceiling.
https://museumsandgalleries.leeds.gov.uk.

THE BREWERY TAPROOMS

There are a few breweries worth visiting on the fringes of the city centre and beyond. **North Brewing**'s Springwell HQ is just off the Meanwood Valley Nature Trail. Opening weekends, with a tap wall pouring the North range, the bar is on the brewhouse floor and spills out into the yard, where pop-up street food stalls provide the solids. Frequent tours and tastings are on offer, and look out for the Springwell Sessions beer and music festivals. info.northbrewing.com/venue/springwell/.

Kirkstall Brewery [8] is out to the west, on the A65, and the tap is more like an on-site pub, spacious, comfortable and decorated with eye-catching breweriana and fittings culled from churches. Beers brewed on site dominate seven lines of cask and a dozen keg taps, but you can also find other Leeds brewers plus a couple from further afield. Brewery tours and tastings are available. kirkstallbrewerytap.co.uk.

Northern Monk Brewery occupies an old warehouse in the Temple district south of the river. Its spacious and bright Refectory is upstairs with 15 lines showcasing beers from some of the best brewers in Britain and Europe, including its own, of course. Look out for one-off collaborations and examples of Northern Monk's Cellar Project experiment with aging dark and wild-fermented brews. www.northernmonk.com/pages/leeds.

On neighbouring industrial estates either side of the A61 at Sheepscar, two small brewers, **Tartarus Beers** and **Bini Brew Co**, have taprooms. The former has 11 lines for sampling an astonishing variety of adventurous beers, from wheat beers and goses to hefty pastry stouts, while the latter's eight lines lead on lagers and pales. www.tartarusbeers.co.uk/pages/tartarus-tap. binibrew.co/pages/taproom.

KEIGHLEY TO SHIPLEY

A handy railway line joins Keighley to the West and Shipley to the East in just 13 minutes, linking some interesting pubs and bars and a few breweries, too. Along the way you can watch barges pass through the steepest flight of locks in the land and visit a model workers' village.

KEIGHLEY

 ### WHERE TO DRINK

Timothy Taylor's Landlord is one of the country's best-selling and most highly regarded cask ales, and while you can find this hoppy golden brew on draught almost anywhere these days, a visit to the home town of the Knowle Spring Brewery ensures you can sample the full **Timothy Taylor** range at its best.

Next to St Andrew's Church, **Taylor's on the Green** [9] acts as a brewery tap. It's a modern, spacious food pub that proudly displays, alongside Landlord, the sweeter Landlord Dark (which used to be known as Ram Tam), the award-winning Boltmaker bitter, the easy-drinking blonde Knowle Spring, and both a dark mild and a light mild – the quaffable Golden Best. There's also a mini-shop on site where you can purchase your Timmy Taylor's merch.

Before the pub opened on the green, the **Boltmakers Arms** was the best place to drink Taylor's beers – and it arguably still is. A small,

cosy local, close to the station, you'll find at least four of the range on the pumps here, all usually in top condition.
www.timothytaylor.co.uk.
www.taylorsonthegreen.co.uk.
www.timothytaylor.co.uk/pubs/boltmaker-arms-keighley.

Also worth a visit while you're in town are the **Brown Cow**, a traditional freehouse offering a well-chosen range of at least six cask ales from across Yorkshire, and the **Red Pig**, a quirkier spot that doubles as a gallery for local artists, and there's usually something interesting on the pumps.
https://thebrowncowkeighley.com.

On an industrial estate the other side of Keighley Station, **Wishbone Brewery** opens its doors for monthly Saturday Sessions during which you can taste a variety of modern styles on the taps created by a former Saltaire brewer.
www.wishbonebrewery.co.uk/.

TIMOTHY TAYLOR'S
TAYLOR'S
ON
THE GREEN
TAP & KITCHEN

[10]

👓 WHAT TO SEE & DO

Keighley, where the River Worth joins the Aire, was once a busy market town and textile centre. A few glimpses of Victorian and Edwardian Keighley survived the overenthusiastic use of the wrecking ball in the 1960s, notably a sweep of buildings along **Cavendish Street** and the first public **library** in England to be funded by the American philanthropist Andrew Carnegie. www.bradford.gov.uk/libraries.

The most interesting diversion is on the outskirts. **Cliffe Castle Museum** [10] is a neo-Gothic pile with a colossal tower above its front door and interiors that reveal the flamboyant taste of the 19th-century industrialist Henry Isaac Butterfield, who moved here after the death of his wife, a member of the Roosevelt clan. Butterfield was a passionate Francophile, and it shows, from the gilded painted plaster ceilings to the fabulous chandeliers and the huge portrait of Napoleon III. https://bradfordmuseums.org/cliffe-castle.

[11]

Alternatively, board a steam (or diesel) train at the town's station and ride the **Keighley & Worth Valley Railway** [11], calling at Oakworth Station, one of the stars of *The Railway Children* films, and Howarth, a town dedicated to its former famous literary residents, the Brontës. In Haworth, you can visit the renowned Brontë Parsonage Museum. https://kwvr.co.uk. www.bronte.org.uk/the-brontes-and-haworth.

BINGLEY

🍺 WHERE TO DRINK

On your way to Bingley, jump off the train at Crossflatts and you're a couple of minutes' walk from **Goose Eye Brewery** [12]. The taproom is upstairs, a large space overlooking the brewhouse where six handpumps pour a changing range of traditional ales.
www.goose-eye-brewery.co.uk.

Opposite Bingley Station, the **Peacock Bar** is a desi pub, sibling to Bradford's Peacock Bar, serving curries, bowls and Indian burritos and wraps, plus an exceptional range of beers, 10 on cask and another four from craft taps.
www.peacockbar.co.uk/peacock-bingley.

On Main Street you've got a great micropub called **Chip n Ern** [13], which has a wacky array of bric-a-brac, including spooky mannequins, and seven cask lines and four keg pouring a good choice showcasing mostly local breweries. An upstairs bar focuses on craft with 10 more taps.
https://chipnern.co.uk.

A few doors down, **Reuben's Beer and Gin House** is a more modern affair, open plan with big windows and a long bar offering a dozen lines of British-brewed craft and continental options and three Yorkshire cask. Keep going and the **Brown Cow** is in a picturesque spot on the banks of the River Aire, a food pub pouring Timothy Taylor's cask beers.
www.reubenshouse.co.uk.
https://browncowbingley.com.

[12]

[13]

👓 WHAT TO SEE & DO

Leave the train at Crossflatts and about a mile along the towpath of the Leeds & Liverpool canal towards Bingley you'll find the **Bingley Five Rise Locks**, the steepest flight in the country, which opened in 1774. It takes about 45 minutes for a barge to pass through the six gates on the 18m (60ft) climb, at the top of which boaters – and onlookers – are rewarded with wide views of the Aire Valley. Closer to Bingley, there's another sequence of three locks in the shadow of the soaring chimney of Bowling Green Mill that now bears the name of the Damart clothing company.
www.bingleywalkersarewelcome.org.uk.
https://canalrivertrust.org.uk/places-to-visit/
    bingley.

For people of a certain age, this town is synonymous with the Bradford & Bingley Building Society which, in different forms, prospered for more than a century until it fell victim to the financial crash in the early 2000s. Don't try to find the original Bingley office – it's underneath Lidl. Instead, take a stroll around the town's historic buildings, including the **Buttercross** on the square and the **old fire station** with its tower peeping above the rooftops where hoses would have been hung out to dry. On the side of the Peacock Bar (see Where to Drink), spot the **tribute to Hindu, Sikh and Muslim pilots** who served in the Second World War.
Pilots tribute, Wellington Street, BD16 2NB.

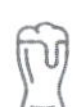

SALTAIRE & SHIPLEY

WHERE TO DRINK

How do you make a World Heritage site an even more attractive destination? How about converting an old tramshed into a spectacular brewery and pub. That's what Ossett Brewery did in 2019, creating the **Salt Beer Factory Bar & Kitchen**. It's a marvellous double space. At the front, there's a large modern pub with a balcony and a horseshoe bar pouring Ossett's cask ales alongside a Salt craft range, which includes a couple of rare Salt casks brewed on the premises. Out back, there's a shiny brewery with its own bar and taproom that extends into the mezzanine. https://taps.saltbeerfactory.co.uk/bar-kitchen.

For contrast, close by on the Bingley Road, you'll find a pair of micropubs. The **Hop Star and Beer Store** has eight keg lines and two cask and a back room hosting live music, while the **Cap and Collar** has a balanced cask and keg offer across six or seven lines and a beer garden. www.facebook.com/Hopstarsaltaire. www.facebook.com/capcollar.

Any beer lover's trip to Saltaire will be incomplete without a visit to **Fanny's Ale House** [14]. Occupying a former police station that retains the blue lamp if little else, this wonderful multiroomed pub sprawls across two floors with an efficiently run bar serving up to nine mostly Yorkshire cask beers with more tempting craft options on the keg lines. www.facebook.com/fannysalehousesaltaire/.

Up the hill, the splendid **Ring O' Bells** is an inviting Edwardian pub packed with architectural features and some decent cask ales, too, at its horseshoe bar, including Tetley's, Saltaire, plus other Yorkshire brews. https://ringobellsshipley.co.uk.

And don't miss **Cultures Deli & Craft House**, a corner café near the station specialising in fermented food and drink, which, of course, includes beer. Expect half a dozen keg lines pouring interesting craft brews to go with your cheese. www.instagram.com/culturessaltaire.

The best way to reach **Saltaire Brewery** and its pleasant taproom is to walk east along the towpath of the canal, which runs parallel to the Aire at Shipley. The brewer is probably most famous for its luscious Triple Choc stout, but with eight craft lines and six cask on site you get a full idea of the range, from single-hop pales to European-style lagers, and punchy IPAs to the low-alcohol ale Northern Light. https://saltairebrewery.com.

Not to be confused with the rural Fox at Shipley – this is the urban variety – **The Fox** is probably the pick of the town centre pubs. Five rotating cask lines typically include some interesting choices from Yorkshire, while six craft taps are shared between continental brews and UK-brewed craft, and the fridges are stocked with up to 50 European ales and lagers. www.thefoxshipley.co.uk.

Other options include the **Crafty Kernel**, a café bar with craft and cask lines offering a busy programme of entertainment, and **Hullabaloo**, a modern-traditional pub with six craft lines and three keg which was in the process of being taken over by the local community as this guide went to press. www.facebook.com/thecraftykernel/. https://hullabalooshipley.org.uk.

WHAT TO SEE & DO

Saltaire is an industrial village and textile mill protected by UNESCO. The industrialist Titus Salt consolidated his production of high-quality woollen cloth, from yarn to finished product, here in the mid 19th-century. Not only did he build a state-of-the-art mill, but also a neat grid of streets of cottages cut from millstone grit for his workers, and facilities including a wash house, hospital and park – all benefits offered on condition of sticking to a strict set of rules. It was a model of urban planning and benevolent dictatorship.

The humble **back-to-backs**, separated at the rear by tight alleys where wheelie bins stand sentry, are now desirable properties that command the attention of the compilers of the *Sunday Times*' 'Best Places To Live Guide'. The looms stopped at **Salts Mill** [15] in 1986 and this vast space now dwarfs an exhibition about Salt's empire and the **1853 Gallery**, which has a permanent display of work by the Bradford-born artist David Hockney. https://saltairevillage.info. www.saltsmill.org.uk.

FURTHER INFORMATION
For more ideas on what to see and do, go to www.visitbradford.com. Check opening times at beer venues and attractions before you go.

[15]

SHEFFIELD

Once Steel City, Sheffield is now Beer City, with a disproportionate number of new breweries driving craft innovation plus a generous sprinkling of traditional pubs doing a brilliant job of bringing good pints to the people. It even has its own Sheffield Beer Report[*], compiled by leading beer writer Pete Brown. You might be surprised that this city has a green heart to discover amid stories about its industrial past.

*Sheffield Beer Report, www.flipsnack.com/uos/sheffield-beer-report-2024.

AROUND THE STATION

 ### WHERE TO DRINK

It begins as soon as you step off the train. **Sheffield Tap** is embedded in the Edwardian station buildings across several panelled rooms, wonderfully restored. The Tap houses its own microbrewery supplying weekly specials and a bar bristling with 11 handpumps for cask and another 14 keg lines. Whether you're waiting for a train, or just got off one, it would be rude not to. https://sheffieldtap.com/.

Then you must go straight to the **Rutland Arms**. Don't argue. It's the law. More than a century old, this traditional pub has retained an unspoilt quirky charm while keeping pace with the exciting developments in beer. Seven handpumps showcase some of the best of modern British brewing while a dozen taps feature continental classics and cutting-edge craft. The graffiti scrawled outside the door says it all for many: 'I love this pub'. www.rutlandarmssheffield.co.uk.

Further to the south-west **Triple Point Brewing** has a spectacular on-site taproom that you enter through an appropriately pointy archway. There's plenty of outdoor seating before you reach a bar serving a full range of the beers you can see fermenting in tanks at the back of the room. Most pour from keg lines and there are usually two or three on cask. Gluten-free beers in many styles are a speciality. Across the road, the **Industry Tap** is a glass-fronted ex-nightclub now dedicated to an impressive range of beers from around the world via 21 keg taps. At last visit, it featured breweries included Verdant, Bianca Road, Full Circle, Duvel, Rodenbach, Vault City, Lervig and New Zealand's 8 Wired. https://triplepointbrewing.co.uk. www.facebook.com/IndustryTapSheffield.

Sheffield's branch of the **Head of Steam** [16] chain is on Tudor Square opposite the Crucible Theatre, best known for hosting the World Snooker Championships. It's housed in a former bank that goes right through to the street at the back and spills out onto tables in the square. Expect the usual bewildering range of cask and keg beers on the bar. www.theheadofsteam.co.uk/bars/sheffield.

Meanwhile, Thornbridge Brewery's £1m beer house The Fargate was due to open opposite the town hall by early 2025.

★ VEDETT ★ ANCHOR BREWING ★ BEAVERTOWN ★
DE HALVE MAAN ★ BROOKLYN BREWERY ★ LA CHOUFFE ★
★ WESTMALLE ★ SIERRA NEVADA ★ OSKAR BLUES ★ MAR
Jupiler
Liefmans
ON THE ROCKS
PALM
LIVE DJ
FRIDAY
SATURDAY
BEERPIG
LUNCH
ABK

[18]

👓 WHAT TO SEE & DO

It may seem an odd idea to send you to a housing estate, yet anyone interested in architecture will want to see **Park Hill**. The Le Corbusier influence is clear in the grid-shaped megablocks, which were built on a hillside behind the railway station in the late 1950s to replace a notorious slum. With the brave new Brutalist world itself becoming a symbol of societal breakdown, a slow and controversial process of renovation into private housing and commercial space began in the early 2000s. Park Hill was always hard to miss for its sheer size, these days it also catches the eye with its multicoloured cladding, aimed at visually distancing the estate from the perception of its bleak past. Work continues as does the debate about it – is Park Hill an eyesore or an important part of Sheffield's heritage, a suitable regeneration or a missed opportunity for social housing? Contemplate it over a coffee at **South Street Kitchen**. www.southstreetkitchen.org.

[17]

CITY CENTRE

 WHERE TO DRINK

Vocation & Co is the Hebden Bridge brewer's Sheffield outpost, a large, bright, U-shaped space with a long bar pouring beers from 22 keg taps and five cask lines. As well as Vocation's own brews, collaborations and special guests feature, most available as flights.
www.vocationbrewery.com/pages/vocation-co-sheffield.

Hidden away in the backstreets to the west of the city centre, the **Bath Hotel** is a beautifully restored 1930s corner house, well worth seeking out for a quiet pint. The island bar serves a lounge, a snug and a drinking corridor, with Yorkshire brews dominating the half-dozen handpumps. Another traditional pub, the **Red Deer** [17], lies a few minutes' walk away on the other side the Glossop Road, readily spotted thanks to a striking mural on its gable end wall. It's a homely sort of place, popular with students and aging regulars alike, and rotating cask ales are well-kept.
https://thebathhotelpub.com.
www.thereddeersheffield.co.uk.

Handily placed close to the famous Butlers Balti House, the **Perch Brewhouse** is the aptly named tap for the city's **Dead Parrot Beer Company**. You'll find a choice of four of its beers on cask, plus guests, including Belgians, on the taps and in the fridge. Then comes the **Crow Inn**, a small, friendly pub with rooms that takes its beer very seriously. There are five ales on cask, featuring Sheffield's own Abbeydale Brewery, plus 10 rotating keg lines that might include classic modern pales from the likes of Deya, rarities from Suffolk's Little Earth Project or Portsmouth's MakeMake, Belgians and sours.
www.facebook.com/PerchBrewhouseSheffield.
www.thecrowinn.co.uk.

Lost in West Bar, on the northern edge of the city centre, is a bar, bottle shop and live acoustic venue in the micropub style, with four handpulls dispensing local ales and a well-chosen range of craft keg on tap. Nearby, **Two & Six** is another micro that's something of a community hub, with an art gallery upstairs that you can hire for events. Some interesting beers, cask and keg, are on offer, too.
Lost in West Bar, 163 Gibraltar Street, S3 8UA.
Two & Six, 26 Snig Hill, S3 8NB.

WHAT TO SEE & DO

Sheffield's claim to be England's greenest city might seem surprising in the centre, where it's hard to see the buildings that survived the 1940 Blitz for those that have filled in the gaps during the past 80 years. Yet, apparently almost two-thirds of Sheffield is parks, gardens and woods, with nearly a fifth of it covered by trees – in fact, it's one of the Tree Cities of the World. As if this were not enough bonding with nature, Sheffield sits on the edge of the Peak District. No wonder the local tourist board calls it 'The Outdoor City' and, if proof is needed, there's the **Greenground Map** with almost half of Sheffield's whopping 800 green spaces plotted on it.

Even in the city centre, where masonry dominates, nature pushes back. The flagstones in the contemplative space of the **Peace Gardens** give way to clipped lawns, cascades and fountains, while the **Winter Garden** nurtures palms, ferns and other exotic plants beneath its graceful high glass arches. A door leads off the Winter Garden to the **Millennium Gallery**, which, along with changing exhibitions and the expected history of the local metalworking industries, contains the **Ruskin Collection**. The minerals, art and books were amassed by the polymath John Ruskin in 1875 to inspire Sheffield's skilled metalworkers to take a break and actively enjoy the beauty of the natural world around them. Just west of the city centre are the **Sheffield Botanical Gardens** [18], which, as well as offering a horticultural tour of the world, have a pit that was once the lair of Bruin the bear, imprisoned for the amusement of Victorian visitors.
www.welcometosheffield.co.uk.
www.sheffield.gov.uk.
www.sheffieldmuseums.org.uk.
www.sbg.org.uk.

THE
FAT CAT

KELHAM ISLAND & NEEPSEND

 ## WHERE TO DRINK

Once a rundown industrial zone, Kelham Island has found a new life as a cool drinking and dining destination – and before it got trendy there was **Kelham Island Brewery** [20]. Founded in 1990 by the late, legendary Dave Wickett, next door to the already famous Fat Cat pub, it was an early champion of US-inspired hop-forward ales in the UK, notably with Pale Rider. Now owned by a local consortium, Kelham Island's beers are still best consumed in the **Fat Cat** [19], which has always served as an eccentric kind of brewery tap, rigorously old-fashioned and dedicated to pouring perfect pints of constantly changing cask beers from its 10 handpumps plus bottled beers from Europe.
www.kelhambrewery.co.uk.
 www.thefatcat.co.uk.

Another historic Sheffield beer experience is available at the **Riverside**, where you can once again quaff a pint of Stones Bitter. Originally designed by Cannon Brewery to quench the thirsts of the city's steel workers, it was at one point the best-selling ale in the country. And now Stones is made in Sheffield again thanks to a deal between **True North Brewing** and brand owner Molson Coors. The Riverside is one of a group of pubs run by True North, in a pleasant spot on the Island with a terrace overlooking the water. Alongside Stones and other ales from the brewer, it has a good choice of guests on cask and keg.
www.riversidesheffield.co.uk.

Kelham Island Tavern is all about the beer. Despite its modest size a dozen handpumps line the bar offering something for everyone, and there's British craft from the likes of Kernel plus continental beers on the taps. Tempting snacks, from pork pies to samosas, are designed with the drinker in mind.
www.kelhamtavern.co.uk.

A much more recent arrival, **Salt Craft + Bao** is the Sheffield outpost of Saltaire's Salt Beer Factory. Housed in a modern, grey-bricked building with a spacious sun-trap terrace out front, it pours the full range of Salt regulars plus specials and also serves Indian street food.
https://taps.saltbeerfactory.co.uk/sheffield.

The pick of neighbouring Neepsend includes two very different venues. **The Wellington** is the **Neepsend Brewery** taphouse, a great little unfussy 'front room' pub with half a dozen handpumps showcasing the local brewer's range, plus a guest or two and a few interesting craft keg lines. **Heist** [21], in contrast, is a vast industrial barn with a brewery, taproom and a tasting lounge for events and beer experiences. As well as an impressive array of its own beers, available in flights and cocktails such as the Lagerita and the Black Forest Espresso Stoutini, there are arcade games. Super fun, if you're in the mood.
www.facebook.com/thewellingtonsheffield.
www.heistbrewco.com.

Even on **Kelham Island**, once bustling with the business of metalworking, green shoots push through along the banks of the **River Don** (you can walk as far as Doncaster if you like), which, with its four tributaries – the Sheaf, Loxley, Rivelin and Porter – fuelled local industry from when the Goit mill race was constructed in medieval times to the advent of steam power. There's even an attempt to lure salmon back here after an absence of 200 years.
https://dcrt.org.uk/about-the-trust/our-rivers/
   the-river-don.

But bricks and mortar dominate Kelham Island and adjoining Neepsend in the shape of the former works and warehouses, which host other creatives alongside the brewers. See artists, designers and makers at work or selling their wares at **Kelham Arcade** and **Peddler Market**, while the multipurpose **Yellow Arch** houses a vast vintage emporium, as well as an events venue and a recording studio where the Arctic Monkeys and Kylie Minogue have made music. There's a foodie scene here, too, best explored with the tasty as well as knowledgeable help of Kelham Island Food Tours.
www.kelhamislandfoodtours.co.uk.
https://kelhamarcade.uk.
www.yellowarch.com.

All these newcomers haven't erased the past; in fact, they've made connections across time, such as the mural outside the **Cutlery Works** [22] food hall showing the local anti-slavery campaigner Mary Anne Rawson with the freed slave and abolitionist Frederick Douglass, who spoke at a rally in Sheffield in 1846. Nearby, **Kelham Island Museum** gives the definitive account of the city's industrial heritage, with exhibits displayed in the impressive setting of a Victorian power station where the 12,000 horsepower River Don Steam Engine steals the show when it cranks into action. Outside the museum is an information board displaying the **Furnace Trail** – snap a photo, as unfortunately it's no longer available as a leaflet. It plots existing industrial buildings and describes a little of their history, giving insights into names such as the Globe Works, so called by its owner because of his quite achievable ambition to sell steel and cutlery around the world.

[22]

https://cutleryworks.co.uk.
www.sheffieldmuseums.org.uk/visit-us/
   kelham-island-museum.

WHILE YOU'RE HERE

There are plenty of fine beer destinations in Sheffield's suburbs, and one in particular is worth an excursion. **The Brewery of St Mars of the Desert** [23] is as quirky and unusual as its name suggests. Best way on foot is to pick up one of the river or canal paths heading east from Kelham Island towards Attercliffe. Hidden among garages and a ball-bearing factory, the brewery's door opens onto a whole new world and a wacky Bavarian apres-ski kind of vibe. Founded by American couple Dann Paquette and Martha Holley-Paquette, SMOD, as it prefers to be abbreviated, specialises in classic continental beer styles (plus an IPA or two) including lagers, saisons and barrel-aged brews. Strange shandies, too, are served in the taproom if you're lucky enough to find a seat. Best book.
https://beerofsmod.co.uk.

FURTHER INFORMATION

For more ideas on what to see and do, go to www.welcometosheffield.co.uk. Check opening times at beer venues and attractions before you go.

Saint Mars
of the
Desert

GUY FAWKES INN

You get a free two-hour guided walking tour with your B&B at The Guy Fawkes Inn in York if you visit Friday to Wednesday. That's on top of a rather attractive room for the night at this medieval hostelry, so called because it's on the site where the Gunpowder plotter was born. The 13 suites will transport you back to those days and are just the place to snuggle up after dark in the old city. Richly decorated in dark shades, crisp white linen is complemented by velvet cushions and thick counterpanes, heavy drapes and solid-wood furnishings, including a four-poster bed (and a view of York Minster) in one room. The look is complemented by all mod cons – in the Belfry Suite there isn't just a TV on the wall, there's a TV lounge with games console. Downstairs at the bar, Saltaire, Black Sheep and Ossett are on the pumps. www.guyfawkesinnyork.com.

MORE PLACES TO STAY

YORK

Dean Court
www.inncollectiongroup.com/dean-court.

The Fat Badger [24]
www.thefatbadgeryork.com.

The Judge's Lodging
https://judgeslodgingyork.co.uk.

The Exhibition [25]
www.exhibitionhotelyork.co.uk.

The Bootham Tavern
www.boothamtavernyork.co.uk.

The Golden Fleece
www.greatukpubs.co.uk/the-golden-fleece-york.

LEEDS

The Calverley Arms [26]
www.innkeeperscollection.co.uk.

KEIGHLEY TO SHIPLEY

The Fleece Inn [27]
Haworth, https://fleeceinnhaworth.com.

SHEFFIELD

The Florentine
www.theflorentinepub.com.

NORTH WEST

Opposite: Black Bull, Coniston.

LAKE DISTRICT

The Lake District has been fertile ground for small breweries serving a busy tourist trade along with the locals, and it's notable for the number of pubs offering accommodation. Its towns and villages are gateways to some of the best walking England has to offer, inspiration for poets and artists down the centuries, but make time to explore their streets, too.

KENDAL

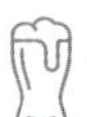 ### WHERE TO DRINK

There's a big emphasis on craft beer here. **Lakes Brewery** is on an industrial estate a short walk from the train station. The taproom is right inside the working brewery so opens only at weekends, pouring the range from a temporary bar. **Gan Yam Brewery** [1] opens its doors even less frequently and is a little further out.
https://lakesbrewco.com/tap-room.
https://ganyambrew.co.uk.

In the centre of town, **Indie Craft Beer** is a small bar and shop over two floors with a dozen taps pouring a variety of styles from top brewers. **Fell Bar** is the eponymous brewery's tap house with a trendy vibe, eight keg lines and six cask featuring Fell beers and guests. And **Handsome Snuff Bar** is Handsome Brewery's house, which peculiarly runs the keg lines over your head and out through a tank on the bar.
www.indiecraftbeer.co.uk.
https://fellbrewery.co.uk/venues/fell-kendal/.
www.handsomebrewery.co.uk.

Across the river, the **Barrel House** [2] is the more conventional tap room of **Bowness Bay Brewing**. There is a mezzanine above the main bar which opens onto a yard with chalet-style huts where you can drink the brewer's beers on cask and keg. The nearby **Factory Tap** is the town's cask beer specialist with nine handpulls mainly showcasing local brewers.
www.bownessbaybrewing.co.uk/the-barrel-house.
http://thefactorytap.co.uk.

On the Windermere Road, the **New Union** is a traditional pub with an extensive beer range taking in four cask and six keg from modern craft breweries, constantly varying, and up to 50 bottled Belgian beers. It hosts tastings and bottle shares to give everyone a chance to sample the rarer brews.
https://thenewunion.co.uk.

WHILE YOU'RE HERE

On the southern edge of the national park at Lowick Green the **Farmer's Arms** is the kind of

[1]

LIVE @ THE
MUSIC
BARREL
HOUSE
Follow Us
The Barrel House
@thebarrelh
CASK
KEG
BARREL
HOUSE
LOWEST DAY BREWERY
[2]

[3]

place you can happily get lost in. Owned by the local community, it's not just a pub, it's a café, an exhibition space, a craft workshop, a live music venue and a place to stay, with two self-catering apartments.
www.hawksheadbrewery.co.uk.
https://lakedistrictfarmersarms.com.

👓 WHAT TO SEE & DO

Don't be too hasty to get to the national park. This busy gateway town has the ruins of a medieval **castle** [4] to see and **Abbot Hall** with its collection of art from Turner to Hockney. Walk along the high street and at every few metres an alley reveals one of the 150 or so yards that developed in the 1800s, from medieval burgage plots to workshops for blacksmiths, coach builders and, of course, brewers. In fact, the clue is in the name of **Brewery Arts** [3], custodian of the town's cultural activities, which is inspired by the former purpose of this building as home to the Whitwell Brewery, which closed in 1968.

Kendal Castle, Sunnyside, LA9 7DJ.
https://lakelandarts.org.uk/abbot-hall.

Such a distinct urban plan should be a prime subject for **Kendal Museum**, but no. Instead, you might be surprised to find amid other local history and artefacts that have washed up from around the world a whole floor devoted to a staggeringly comprehensive collection of antique taxidermy. But then Kendal is full of little surprises, such as the **Old Post Office** on Greenside, now marked by a plaque, which inspired the late former resident John Cunliffe to write Postman Pat, and the extraordinary tapestry on display in the **Quaker's meeting hall**, created in the 1990s by members of the faith around the world. The embroidered panels are as thought-provoking as they are beautiful, depicting the Quakers' wide range of interests and endeavours, from botany to the abolition of slavery.
https://kendalmuseum.org.uk.
The Old Post Office, 10 Greenside, LA9 4LD.
www.quaker-tapestry.co.uk.

AMBLESIDE & GRASMERE

🍺 WHERE TO DRINK

At the centre of the tourist hub of Ambleside there is, surprisingly, a tank beer bar. **Tap Yard** [5] is on the site of an old mill and features six copper tanks suspended above the bar from which beers from various brewers are poured via a network of pipes. Handpulled cask ales are also available, and the pizzas are very popular with families. www.tap-yard.com.

The neighbouring village of Grasmere is a relative haven of calm and has two excellent pubs. The **Good Sport** is the tap house for Grasmere Brewery & Distillery, 100 metres away, and serves its range on cask and keg. **Tweedies Bar & Lodge** is one of those rare hotels that's truly serious about beer and offers a great selection from handpump and tap in a comfortable bar and dining room that spills out into a courtyard garden. www.grasmerepub.com. www.tweediesgrasmere.com/tweedies-bar.

A winding road out of Grasmere takes you to the **Britannia Inn**, perched in a beautiful location in the tiny village of Elterwater. Its owners also run Langdale Brewery so you'll always find well-kept examples of those beers on the pumps as well as other local brews. www.thebritanniainn.com.

🔭 WHAT TO SEE & DO

Follow the road to **Stock Ghyll Force** and you'll find the reason why **Ambleside** exists. As you climb the wooded path, the neighbouring stream races past in the other direction, powered by a 21m (70ft)-high waterfall 800m (0.5 miles) up. The swirling waters charge down through town, passing beneath the tiny 17th-century **Bridge House**, which we must believe a family of eight once lived in, to the River Rothay.

For centuries, it powered the mills built on its banks, which are occupied today by shops and restaurants.
Stock Ghyll Force, Stock Ghyll Lane, LA22 0QT. www.nationaltrust.org.uk.

This natural phenomenon was also one of the designated viewpoints, or 'stations', to be admired according to Thomas West, author of *Guide to the Lakes*. The first book of its kind about the area, published in 1778, it helped trigger a tourism boom, aided by the fading popularity of the Grand Tour largely thanks to the Napoleonic Wars. Ambleside became an essential stop on a trip around the area, close to England's largest lake, **Windermere**, which we still can't resist taking cruises on.
www.windermere-lakecruises.co.uk.

Artists, poets and writers have continued the easy sell of this picturesque pocket of hills, valleys and lakes, Turner, Constable, Wordsworth and Potter among them. But what about Kurt Schwitters? The German artist and poet moved to Ambleside in 1945 and created the Merz Barn in nearby Elterwater, his studio until his death in 1948. The small stone building is regarded by some as one of the most important Modernist works in Britain but closed to the public in 2022. You'll have to satisfy yourself with seeing Schwitters' bust outside the **Armitt Library and Museum**, which sometimes displays his paintings of the area.
www.armitt.com.

The tradition of capturing visions of the local landscape continues at the **Heaton Cooper Studio** in Grasmere where the artist William Heaton Cooper settled in the 1930s. Today his watercolours are on show alongside work by family members in a bright modern gallery that strikes a contrast with the surrounding grey slate buildings sheltering beneath the fells. But William Wordsworth is the main attraction here. Walk around the village and the theme becomes apparent. The National Trust's villa Allan Bank was once his home, **St Oswald's church** with its unusual wood-beamed ceiling is where he worshipped and his grave can be found, and the adjacent daffodil garden was planted in his honour.
www.heatoncooper.co.uk.

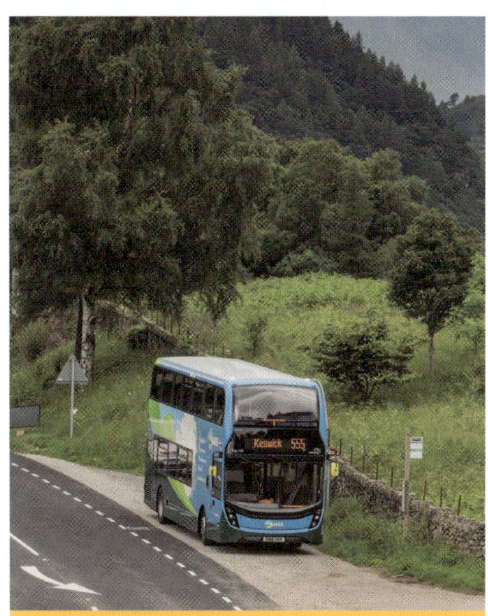

GO SLOW

Our slice of the Lake District is easy to see on public transport. **Aviva West Coast** trains serve Oxenholme, where you can connect to Kendal on the local **Northern** service, and Penrith, where you can catch the **Stagecoach** X5 bus to Keswick. Stagecoach bus 555 provides a regular link between Kendal and Keswick, calling at Ambleside and Grasmere, and between Bowness and Grasmere you can hop on the Lakesider 599, an open-top bus with an audio commentary about what lies around you. In summer 2024, Avanti West Coast and Stagecoach launched a combined rail and bus ticket that takes you through to the town, just select 'Keswick station'.
www.avantiwestcoast.co.uk.
www.stagecoachbus.com.
www.northernrailway.co.uk

WHILE YOU'RE HERE

In 2024 *Bluebird K7*, the jet-engine hydroplane in which Donald Campbell notched up seven world water-speed records, triumphantly returned to Coniston. He died on his eighth attempt on the lake here. His achievements are celebrated in a delightful little museum in the centre of the village. https://ruskinmuseum.com.

STAY AT A BREWPUB

Here are a few of the many pubs in the Lakes that not only have a brewery on site, but also offer rooms.

For cask beer devotees the **Black Bull Inn** at Coniston makes for an enticing pilgrimage. In 1998 it produced the surprise winner of Champion Beer of Britain. Coniston Bluebird, named after Donald Campbell's record-breaking hydroplane (see page 49), was a pioneer of pale hoppy ales and there's an extra frisson in drinking it on the spot it was born. The characterful main building has four double rooms, a twin and a family room, and you can also stay in the Old Man Cottage and the Bluebird Cottage next door.
https://blackbullconiston.co.uk.

The remote **Kirkstile Inn** by Loweswater is the original home of Cumbrian Ales, not brewed at the pub any more since success forced its move to a bigger premises near Hawkshead. But it's still a great pub and every visit feels like an adventure as you sit there with your pint of Loweswater Gold. It has 11 rooms, seven of them in the main building, including two large suites in the annex.
Kirkstile Inn, www.kirkstile.com.

Barngates Brewery is based at the **Drunken Duck Inn** [6], a smart modern conversion with an unusually broad range of beers brewed on the premises including a choice of lagers and a weissbier on keg. It has nine rooms of various standards and sizes plus two suites.
https://drunkenduckinn.co.uk.

The Watermill Inn at Ings, just up the road from Hawkshead Brewery on the River Gowan, is a cosy, picturesque country brewpub and hotel with eight doubles, a single and a family room, that's home to Windermere Brewing Co.
https://lakelandpub.co.uk.

MANCHESTER

Manchester has been at the vibrant centre of the UK's burgeoning craft beer scene for more than a decade, attracting some great brewing talents who have added a fresh dimension to the city's industrial heritage. What also sets it apart, though, is the survival of traditional family brewers, providing a well-rounded experience. There's plenty to see here between the pubs, but picking a theme such as equal rights can offer a new perspective.

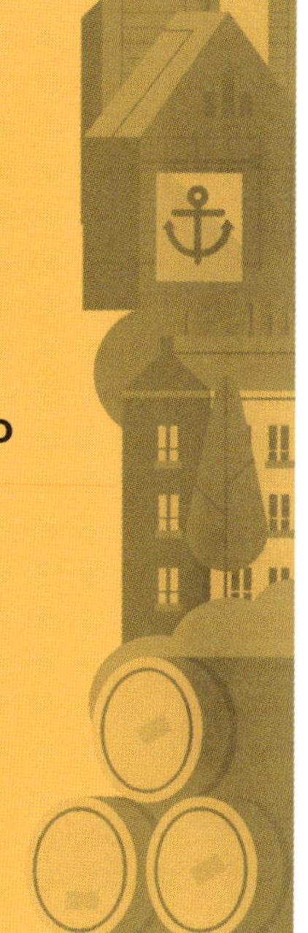

PICCADILLY, GAY VILLAGE & SPINNINGFIELDS

WHERE TO DRINK

The arrival of **Cloudwater Brew Co** in 2014 marked something of a watershed in British brewing, a company that had the highest ambitions not just for its beers but its entire ethical approach – and it happened in Manchester. Its Unit 9 Taproom is next door to the brewery on a trading estate behind Piccadilly Station, where it occupies a pleasant mezzanine area that looks out over the brewer's barrel-aging store. A couple of dozen lines give a sense of the huge range of styles Cloudwater has explored, from low-alcohol pales to sours to pastry stouts.

Of arguably equal stature, **Track Brewing** is on the same estate, just across the way. Its large, welcoming taproom is on the brewery floor and opens out onto a beer garden. Families especially like it. The bar pours 20 keg lines and three cask including the flagship Sonoma pale in both forms. Brewery tours available.
www.cloudwaterbrew.co.
www.trackbrewing.co.

Under the railway arches, **Balance Brewing & Blending** offers something different. It specialises in mixed fermentation beers, made with wild yeasts and matured in oak barrels. Its taproom has four rotating draught lines showcasing the latest expressions of the style. Tours and tastings can help you get a handle on this fascinating brewing process. On the same stretch, the fourth taproom in this stellar cluster is the home of **Sureshot Brewing**. A dozen of its beers are poured here, dominated by IPAs with wacky names. It's a fun take on what can be the very serious business of beer.
https://balancebrewing.co.
https://sureshotbrew.com.

Out the front of the station, **Piccadilly Tap** is the spot for one last third or picking up a can for the train. There's a choice of 17 on the keg lines and another six on cask, representing the diversity of British and continental brewing, served from a curious L-shaped counter that extends through the middle of the pub like a catwalk for bartenders.
www.piccadillytap.com.

At the top of Piccadilly in a basement next to Wetherspoon's you'll find the first of Manchester's two **Bundobust** restaurants. As you hit the bottom of the stairs it opens into a surprisingly large and lively space with a bar at the end serving its own-brewed beers to accompany your Indian street food. Best to book a table for this one. www.bundobust.com/locations/manchester.

Loudly announcing itself with a fabulous mural just off the famous Canal Street circuit in the Gay Village, the **Molly House** is split over two floors. The pubbier part is on the ground floor where up to five well-kept local cask ales are served alongside a few craft selections. www.themollyhouse.com.

Heading west you'll find a pair of good pubs where you can sample the wares of two of Manchester's family brewers. The **Old Monkey** occupies a single room and the bar showcases Joseph Holt cask ales on six pumps plus the results of its recent foray into craft keg. The **Grey Horse Inn** does the same for Hydes Brewery in a more traditional bijoux setting. And squeezed between them like the dormouse at the Mad Hatters tea party is the **Circus Tavern**. Billed as the smallest pub in Europe (but it probably isn't), it consists of a tiny bar with two handpumps and an adjoining room. It's certainly characterful and the beer is well kept – go for the Tetley's if it's on. www.joseph-holt.com/pubs/old-monkey. www.hydesbrewery.com/venue/grey-horse/. https://circustavern.co.uk.

Sticking with tradition, the **City Arms** is one of the city's great heritage houses. The front bar pours up to eight cask ales, including modern styles, and a few interesting craft beers, too, while the room at the back is a great place to hide away. It's one of those pubs that's hard to leave. Opened in 2024 opposite Manchester Town Hall, **Founder's Hall** is a celebration of JW Lees Brewery and a sign of how far this 200-year-old family firm has come.

Alongside the traditional cask ales you'll find the latest limited edition Boilerhouse craft beers – expect the experimental.
www.facebook.com/cityarmsmcr.
www.foundershall.co.uk.

Keep going and before you hit Salford drop into the **Gas Lamp** [7], a wonderful basement pub run by another of Manchester's exciting craft brewers, Pomona Island. Comprising two rooms with a corridor between, it's on the site of a former charity soup kitchen and the walls are lined with cream tiles. Not your usual pub decor. Three or four Pomona beers are served on cask while 10 keg lines give a sense of the brewer's range. A gose, a barley wine and a witbier were all on at last visit.
www.thegaslampmanchester.co.uk.

Back towards the centre, in a former bank building, Pomona Island also has **North Westward Ho**, more traditional in style, high ceilinged and expansive, displaying 18 craft lines featuring many styles plus five ales on cask. Practically next door is another basement venue, **Sam's Chop House**, a quirky pub-restaurant where the waiters still wear aprons and glide noiselessly between the tables. LS Lowry was reputedly a regular here and a life-sized bronze statue of the artist props up the bar. If you're not too creeped out by that, it's a comfy spot to enjoy a pint from that other local legend John Willie Lees.
www.northwestwardho.co.uk.
www.samschophouse.com.

If there's one place that stands for Manchester's eclectic beer scene it's **Café Beermoth** [8], a big, open, glass-walled space lined with booths and a long bar pouring 10 keg lines and seven cask, all carefully chosen to represent the best of British brewing, plus some special treats in bottle and can. Guided tastings by the knowledgeable staff are on offer to small groups.
www.beermoth.co.uk/cafe.

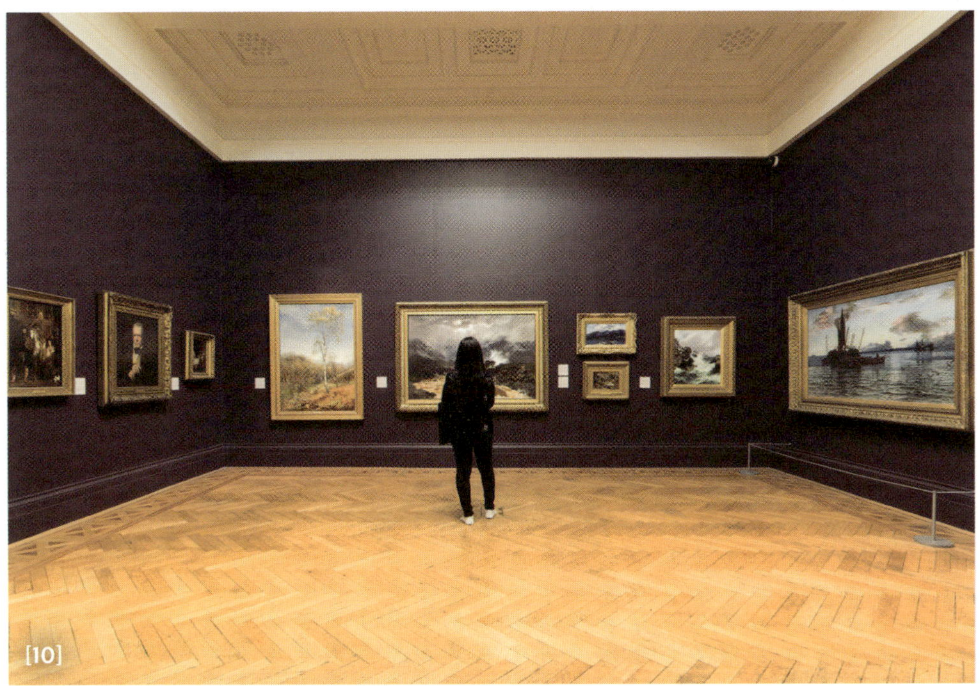

WHAT TO SEE & DO

Pick any theme and you'll find a way of using it to explore Manchester. But one theme that weaves right through this radical city's fabric is equal rights. Step into the Gay Village and the issue is immediately raised by the sight of the sculpture of **Alan Turing**, the mathematician, computer scientist and codebreaker who was persecuted for his homosexuality. At St Peter's Square, the suffragette **Emmeline Pankhurst** [9] is captured in the bronze *Rise up, Women*, arguing in perpetuity for women's right to vote, and a plaque on the **Free Trade Hall** (now The Edwardian Manchester hotel) notes that it was built on the site of the **Peterloo Massacre**, where an anti-poverty protest was attacked by soldiers on 16 August 1819. A **memorial** by the Turner prize-winning artist Jeremy Deller, inscribed with the names of the 18 people who died at the scene, stands nearby on Windmill Street. The Free Trade Hall is one of the venues where the freed slave Frederick Douglass campaigned for the abolition of slavery, and the **statue of Abraham Lincoln** commemorates the American president's tribute to the Lancashire people for boycotting cotton picked by enslaved people in America.

Alan Turing Memorial, Sackville Gardens, M1 3WA. Emmeline Pankhurst statue, and the Peterloo plaque, St Peter's Square, M2 3AE. www.peterloomassacre.org. Abraham Lincoln statue, Lincoln Grove, M2 5LF.

As at many major cultural institutions across the country, a major rethink of how exhibits are presented has taken place at the **Manchester Art Gallery** [10], in particular its Grand Tour gallery and the colonial and racist attitudes that led to artefacts from the Middle East and Asia winding up in this rainy English city. All these threads of the battle for equality are knitted together at the excellent **People's History Museum**, a campaigning institution that champions the fight for a fairer society and the people engaged in those struggles. Its exhibitions delve into the museum's growing archives of objects, pamphlets, papers, photos, posters and banners, from the Chartists to the Miners' Strike, touching on every aspect of working-class action, from trade union rights to the climate emergency and resistance to anti-protest legislation. https://manchesterartgallery.org. https://phm.org.uk.

[11]

NORTHERN QUARTER & VICTORIA

🍺 WHERE TO DRINK

Port Street Beer House was a pioneer of the modern beer age when it opened in 2011, and it's still a leading venue with 18 keg lines and seven cask pouring a fantastic range from brewers in Manchester and further afield in a relaxed atmosphere over two floors. It also hosts some interesting events for those who want to improve their understanding of beer. Nearby, **Northern Monk** [11] has a sizable refectory with three rooms giving it plenty of space for events. The main bar is a stripped-back bare boards and bricks kind of place with 18 taps pouring mostly the Leeds brewer's beers. In 2024 Cumbria's Fell Brewery took over the well-known **Pelican** craft bar, promising a revamp. www.portstreetbeerhouse.co.uk. www.northernmonk.com/pages/manchester. pelicannq.com

Edinburgh's **Fierce Beer** also has a site in the Northern Quarter over two floors plus a terrace in front. Fifteen keg lines give a good idea of its range, which includes sours among the IPAs, and there's usually room for a couple of guests. **MicroBar** is an odd little place inside Arndale Food Market. A few cask ales are sold alongside a host of bottles. www.fiercebeer.com/pages/manchester-bar. www.facebook.com/manchestermicrobar.

In the other direction, by the ring road, the city's **Blackjack Brewing** runs two contrasting outlets close to each other yet in very different surroundings. **Smithfield Market Tavern** is a modern-traditional pub with six cask lines and 12 keg pouring the Blackjack range plus a guest or two, while **Jack in the Box** is a bar inside the Mackie Mayor food court offering a similar selection, the beer 'cellar' being a giant fridge behind the bar. Smithfield Market Tavern, 37 Swan Street, M4 5JZ. www.facebook.com/JITBMackieMayor.

[12]

Sibling venue to the Piccadilly Tap, the **Victoria Tap** is right inside Victoria Station and a great example of a modern railway bar. Its 19 keg lines and five cask are displayed on a departures board above the bar and feature trusted names from British craft brewing alongside interesting continental lagers. Nearby, in the Sadler's Yard development, Cloudwater's **Sadlers Cat** brings the beers. Fourteen lines pour from the tap wall and you can also drink three of the craft brewer's ales on cask, which is a bit of a treat.
Victoria Tap, Victoria Station, M3 1WY.
www.sadlers-cat.pub.

It's a little outside the city centre but there's no way we're not going to give one of Manchester's best pubs a mention. **Marble Arch Inn** [12] was the original home of Marble Beers and still functions as the brewer's public face. It's a beautiful building, Grade II-listed, pink stone on the outside and inside a mosaic floor that cunningly slopes down to the bar, and a simply stunning ceiling. Nine handpumps and eight keg taps serve Marble's beers and a few well-chosen guests.
www.marblebeers.com/the-marble-arch.

WHAT TO SEE & DO

Home to two of the world's top teams, Manchester is a fitting location for the **National Football Museum**. Enjoy the chance to take a shot in a virtual penalty shoot-out and to commentate on *Match of the Day*, exhibitions regularly tell the story of the women's game and tackle the issues of racism and homophobia in the sport. Across the road, in **Chetham's Library**, you can see the desk where Karl Marx and Friedrich Engels studied, Engels researching his work *The Condition of the Working Class in England*. The pair feature in one of the **mosaics** by Mark Kennedy on a wall in the Northern Quarter, along with Pankhurst, Turing and others. Look out for Kennedy's rainbow mosaics on pavements around the city, part of an old gay rights heritage trail, superseded by the current **LGBTQ+ Heritage Trail**.
www.nationalfootballmuseum.com.
https://library.chethams.com.
Mosaic art, Tib Street, M4 1PW.
www.visitmanchester.com/ideas-and-
    inspiration/lgbt/everyone-welcome-
    manchester-lgbtq-walking-trail.

OXFORD ROAD

 ## WHERE TO DRINK

Alongside the canal, with a waterside terrace, **Rain Bar** [13] is an imposing JW Lees house, showcasing the brewer's full range in cask and keg. And a hop across the road will take you to one of Manchester's most famous pubs, the Grade II-listed **Peveril of the Peak** [14], with its ornate exterior and decorative stained-glass interior windows. Choose from two bars to sit in and from four locally brewed cask ales to drink. Another historic pub, the **Britons Protection**, is nearby and displays even more splendid architectural features – including its toilets. Eight handpulls deliver the beer and there's an extensive whisky list, too. www.rain-bar.co.uk.
Peveril of the Peak, 127 Great Bridgewater Street, M1 5JQ.
www.facebook.com/thebritonsprotection.

In contrast, **Vocation & Co.**, operated by the eponymous Hebden Bridge brewer, is about as modern as it gets. It's the hub of a development called Society, which brings together a variety of street-food vendors on the ground floor of an office block by the Rochdale Canal. Most of the 35 lines pour Vocation beers in a variety of styles, and there's room for guests from Deya, Kernel and Sureshot. www.vocationbrewery.com/pages/vocation-co-manchester.

Bundobust Brewery is next door to the Palace Theatre. This is where the beer for the Indian street-food chain comes from and the brewhouse adjoins a large taproom-restaurant where you can sample the excellent range, including lagers, pales, porters and more, with usually at least one on cask. On the other side of Oxford Road Station, in the mind-bogglingly named Circle Square development, you'll find the **North Brewing** taproom pouring the Leeds brewer's range including tempting specials.
https://bundobust.com/locations/bundobust-brewery.
https://info.northbrewing.com/venue/north-taproom-circle-square.

WILSONS
PEVERIL OF THE PEAK
BAR FOOD
BAR GAMES
Wines & Spirits

South of Oxford Road Station the equal rights trail continues. **Manchester Museum** reopened in 2023 after a two-year £15 million revamp. Once, the act of gathering the 4.5 million objects beneath the roof of this splendid neo-Gothic institution, designed by Alfred Waterhouse, architect of the Natural History Museum in London, was of little importance to its guardians. Today, the museum has its first Curator of Indigenous Perspectives and among the fresh displays is the South Asia Gallery, curated by Nusrat Ahmed in consultation with members of the Asian community. Objects such as an opium pipe confront the actions of British imperialism, while original works celebrate the region's creativity. The richly decorated cycle rickshaw was designed by traditional rickshaw artist Sayed Ahmed Hossain from Bangladesh with three British Asian artists.
www.museum.manchester.ac.uk.

Further along, **The Pankhurst Centre** tells the story of the suffragette movement in the home of Emmeline, Christabel and Sylvia Pankhurst, where the first meeting of the Women's Social and Political Union took place in 1903. Around the corner, the final home of **Elizabeth Gaskell** also doubles as a museum about the author and her writings about working-class life in the industrial North. As you're here, there's bound to be something interesting to see at **The Whitworth** art gallery.
www.pankhurstmuseum.com.
https://elizabethgaskellhouse.co.uk.
www.whitworth.manchester.ac.uk.

FURTHER INFORMATION

For more ideas on what to see and do, go to www.visitmanchester.com. Check opening times of beer venues and attractions before you go.

The Castlefield Viaduct: Nature thrives where trains once passed over this Victorian viaduct. The bridge, built by the same engineers who constructed Blackpool Tower, is now a sky garden for everyone to enjoy. www.nationaltrust.org.uk/visit/cheshire-greater-manchester/castlefield-viaduct.

LIVERPOOL

Liverpool enjoys some of the finest pub architecture in the country and some of its most characterful pubs serving great cask ale. A few distinctive craft brewers complete the scene. For visitors, the focus is on the waterfront, a veritable museum park, but save a little time to wander into the city centre, too.

 ## WHERE TO DRINK

Liverpool Brewing rightly has its tap house at the city's hub, between Lime Street and Central stations and just across the road from the landmark Vines pub. Eight craft lines and six cask showcase its own brews plus a sprinkling of guests, for consumption in a relaxed atmosphere over three floors. It's not been dubbed 'the Sanctuary Bar' for nothing. Deep into shopping country, the **Head of Steam** offers further sanctuary, along with the great range of cask and continental beers you expect from the chain. Offering the usual tempting combination of Indian street food and its own range of craft beers, **Bundobust** is hard to spot, up a flight of stairs, opposite the rather more prominent Albert's Schloss, a spectacular, glittering beer hall featuring continental classics on draught alongside a couple of UL-brewed craft selections. www.liverpoolbrewingcompany.com/. www.theheadofsteam.co.uk/bars/liverpool. bundobust.com/locations/liverpool/.

As the name suggests, the **Bridewell** [15] occupies a former jailhouse. Entering the Grade II-listed Victorian building via a gateway and courtyard you'll find the 'cells' have been wittily converted into seating areas and some original fittings retained to create a sense of its history. Four cask lines lead on Kirkstall Brewery ales and the keg taps pour a few continental classics and craft beers. https://thebridewellpub.co.uk.

Hard to believe the **Red Lion** became a pub only in 2022. It's got a traditional feel thanks to a thoughtfully structured main bar area with handy shelving for your glass and a decor dotted with stuffed animals in cases. There's a pleasant courtyard out the back, too. Five cask ales are on offer, with Ossett and Salopian typically among the choices. www.redlionpubliverpool.com.

Heading towards the university, the **Dispensary** is one of Liverpool's famous watering holes. Half a dozen pumps pour cask while there are also a few craft keg taps these days, including a choice of alcohol-frees. The even more famous **Roscoe Head** around the corner is one of only a handful of pubs that have appeared in every edition of the Campaign for Real Ales' *Good Beer Guide*. It's an unspoilt gem with a reverence for cask ale and four homely areas to sit and revere. Nearby, **The Grapes** in Roscoe Street (one of two with that name in the city) focuses on modern styles from its nine handpumps, sourced from craft brewers around the UK. www.facebook.com/Dispensarybydockleaf. www.roscoehead.co.uk/. The Grapes, 25 Mathew Street, L2 6RE.

THE
BRIDEWELL
CAMPBELL
SQUARE

If you don't believe a pub can be breathtakingly beautiful you've obviously never been to the **Lion Tavern**. On the corner next to Moorfields Station, the restored Art Nouveau interior dates from the early 20th century. Behind the main bar the corridor that joins two more drinking areas is tiled in Art Nouveau style with etched glass windows above framing the hatch where you order your beer from a choice of eight cask pumps. The **Denbigh Castle** is also worth a visit. Sibling pub to the Bridewell, it offers four cask lines and five craft plus continental lagers.
https://theliontavern.co.uk/.
www.thedenbighcastle.co.uk.

Dead Crafty is the city's craft beer emporium. A bustling corner bar over two floors, 20 lines feature some of the best of modern brewing in the UK including IPAs, lagers, stouts and sours from producers such as Pomona Island, Beak, Polly's and Pretty Decent, all served by a knowledgeable team.
www.deadcraftybeercompany.com.

Across the busy junction, you'll spot the pale grey Art Deco frontage of the **Ship & Mitre** with, allegedly, the biggest range of beers in Liverpool. The central bar pours around 40 draught lines including eight on cask featuring modern styles and a dozen British craft alongside Belgians and a great selection of world lagers and wheat beers. And if that's not enough, there's plenty more in the fridges.
www.theshipandmitre.uk.

Housed over two floors of a wedge-shaped Grade II-listed building opposite Wapping Dock and the Convention Centre, the **Baltic Fleet** is another celebrated Liverpool pub. Nine lines of cask and keg showcase a variety of styles from local brewers in particular, including the city's own Neptune and Black Lodge.
www.balticfleet.co.uk.

Black Lodge Brewing itself is hidden away in the backstreets with a cosy taproom among the vessels serving 10 lines of craft from pales and pilsners to stouts and sours – and at last visit a surprise best bitter on cask! A sprawling yard out front hosts street-food stalls. Beer tastings with or without food pairing are available.
www.blacklodgebrewing.co.uk.

Love Lane Brewery [16] has had its ups and downs but it's a splendid space brewing its own craft beers and distilling gin, too, within sight of the high-ceilinged bar and restaurant at the front. It owns the Higson's brand, keeping alive the name of Liverpool's famous old brewery, which you can still see to the south of here.
www.lovelanebrewery.com.

WHILE YOU'RE HERE

Head out past the Liver Birds to a small industrial estate by Princes Dock and you'll find a pair of Liverpool's craft brewers, both with taprooms. Behind a small brightly painted door, **Carnival Brewing**'s brewhouse and tasting room offers a choice of pales, IPAs and stouts on keg and a couple on cask. **Azvex Brewing** is a little further on, pouring its somewhat edgier range of beers from a dozen taps. Both provide brewery tours, giving an insight into the brewing process and the secrets of different styles.
www.carnivalbrewing.me.
www.azvexbrewing.com.

WHAT TO SEE & DO

It's easy for those who don't live there to forget that Liverpool was once a major global gateway. The port city's pivotal role in importing and exporting goods, and the clever interconnected dock system, earned Liverpool recognition by UNESCO in 2004. That was snatched away in 2021 due to development that UNESCO didn't like; bold contemporary structures have popped up on **Liverpool's waterfront** [17] in the past 20 years alongside the historic warehouses that were revamped in the 1980s. Whether or not you approve, there's a vast riverside park of museums and galleries at your disposal, and it's all largely free.

Before you step inside, stop for a moment to take in the magnificent backdrop to all this. On one side flows the River Mersey (catch the famous ferry) and on the other stand the Three Graces – the **Port of Liverpool Building**, the **Royal Liver Building** and the **Cunard Building** – with the **Liver Birds** on their high perches. You can find out more about how the harbour made this city's fortunes in the **Merseyside Maritime Museum**, which includes the **International Slavery Museum**, where Liverpool goes further than most in owning its role in the heinous trade in human beings. As well as dealing with the brutal realities of the transatlantic slave trade and reflecting on modern slavery, the museum gives space to displays about life before slavery in West African communities. The **Museum of Liverpool** lightens the mood and expands the view from the docks to the city, tracing Liverpool's roots. Its gallery The People's Republic is particularly entertaining, celebrating the Liverpudlian spirit and the diverse communities, ethnically and spiritually, that have congregated here.
www.liverpoolmuseums.org.uk/maritime-museum.
www.liverpoolmuseums.org.uk/international-
    slavery-museum.
www.liverpoolmuseums.org.uk/museum-of-
    liverpool.

There's art here, too. **Tate Liverpool** [18] is due to reopen in 2025 following a two-year, £30 million revamp, and **Open Eye** shows innovative

[17]

TATE
TAT
Igo
ondinone
iverpool
Mountain

[19]

photography. Sculptures and memorials dot the quay, including Ugo Rondinone's colourful and gravity-defying rock pyramid *Liverpool Mountain*, and the Superlambanana, Taro Chiezo's fusion of a sheep and a banana, once common cargo here, a warning about the dangers of genetic engineering (the original is in Tithebarn Street). But the statue the crowds gather around is of The Beatles [19], another global export. **The Beatles Story Museum** is on the waterfront, and the Fab Four star alongside what came before and after across the road at the **British Music Experience**, a nostalgic romp through the 20th and 21st centuries. There are cases stuffed with memorabilia, superstars in concert on big screens, videos of hitmakers analysing different musical genres, from Merseybeat to grime, and the chance to unleash your inner Jimi Hendrix on a Gibson Les Paul.
www.tate.org.uk/visit/tate-liverpool.
https://openeye.org.uk.
www.beatlesstory.com.
www.britishmusicexperience.com.

You'll find the venue where The Beatles made their name, the **Cavern Club**, in the city centre on Mathew Street, which still hosts live music in part of the original site. After all, it's time to see something of Liverpool, away from the waterfront. You've still to visit the two **cathedrals** and wander the cobbled streets of the **Georgian Quarter** between them, step into **Chinatown** through the largest Chinese Arch outside China, and you might catch a craft fair in **St Luke's Bombed Out Church**. There are treasures to admire in the **Walker** and **Bluecoat** galleries, spectacular interiors to goggle at in the Grade I-listed neoclassical **St George's Hall**, and you could take a selfie in the trendy Baltic Triangle with Liverpool FC legend **Jürgen Klopp**. Well, his mural, at least.
www.cavernclub.com.
https://liverpoolcathedral.org.uk.
https://liverpoolmetrocathedral.org.uk.
www.slboc.com.
www.liverpoolmuseums.org.uk/walker-art-gallery.
www.thebluecoat.org.uk.
www.stgeorgeshallliverpool.co.uk.

FURTHER INFORMATION

For more ideas on what to see and do go to www.visitliverpool.com. Check opening times of beer venues and attractions before you go.

Even those who don't drink beer can get a lot out of a pub crawl of Liverpool. There are some wonderful pieces of architecture and quirky design, too.

No visit to the city should be complete without calling in at the **Philharmonic Dining Rooms** [20], to give it its full name. For some reason, the Phil is best known for its toilets, but the rest of this grand, multiroomed palace of a pub over five floors is stunning, too. And just look at those golden Art Nouveau gates framing the entrance. www.nicholsonspubs.co.uk.

Sharing the same architect, and known to locals as the 'Big House', the **Vines** [21] has been restored to its Edwardian baroque glory. It's completely over-the-top but if you can't do that in a pub, where can you? Don't miss the huge lounge at the back with its domed skylight. www.vinesbighouse.co.uk/.

Originally Pearl Assurance offices, **Doctor Duncan's** dates from a slightly earlier period and exhibits a riot of colourful ornate tiling along with some interesting medical bric-a-brac to celebrate the nation's first Medical Health Officer. www.doctorduncansliverpool.com/.

Over in the city's Georgian Quarter is the work of the eponymous landlord **Peter Kavanagh** who made his idiosyncratic mark here in the first half of the 20th century. Lots of features and fittings to explore here from a **hotch-potch of periods**. Peter Kavanagh's, 2–6 Egerton Street, L8 7LY.

CHESTER

Chester might not be the first place that leaps to mind when you fancy a beer break but it offers a great variety of venues, covering cask and craft. It's also a beauty of a city, layered with history from Roman times onwards yet kept relevant by its university and a lively cultural scene. Just follow the ancient walls and 'Rows'.

 ## WHERE TO DRINK

First stop out of the station has to be the **Deva Tap**, housed in an unusual piece of Arts and Crafts architecture that used to be the Railway Cocoa Rooms, a temperance initiative to compete with pubs. Now it's a great champion of beer with at least six lines of craft keg and three cask. Looks like pubs won the contest. https://thedevatap.co.uk.

On the canal beside the City Road Bridge, the **Old Harkers Arms** is the pubbier of two Brunning & Price houses here, combining good food with half a dozen well-kept local ales on the pumps. Nearby, **The Cellar** has a rustic ale house feel about it with six cask lines, including modern styles and 12 genuinely international keg taps featuring classic American IPAs, continental lagers and Belgian lambics. It hosts a monthly tutored tasting titled 'Beer Without Fear'. Its sibling **The Cornerhouse** is bright and contemporary and more about the food, but it still offers four cask lines and 10 keg. www.brunningandprice.co.uk/harkers. www.thecellarchester.co.uk. www.cornerhousechester.com.

That Beer Place has expanded from a market stall into a smart and welcoming bottle shop and bar on Foregate Street. Split over two floors, it has 11 taps offering a great variety of craft styles, including wild fermented ales, and at least 350 bottles and cans pack the fridges. Friendly and knowledgeable staff are there to talk you through them and there are regular tastings and meet-the-brewer events, too. On the next street, the Chester branch of **Brewhouse & Kitchen** [22] offers a range of cask and craft beers brewed on the premises, while **Spookton Brew Co** unveiled its new brewery and taproom nearby in 2024. https://thatbeerplace.co.uk. www.brewhouseandkitchen.com/venue/chester-2. www.spookton.co.uk.

On the main road north of the Old Dee Bridge, three tempting cask beer venues vie for your custom. The **Bear & Billet** is one of the city's most famous old pubs, spread over three storeys of a Grade I-listed half-timbered building that dates back to the 17th century. Five cask ales are served. Opposite, the **Cross Keys** is a Joule's Brewery housed in a beautifully restored Victorian building with plenty of cosy places to sit amid the panelled oak and stained-glass windows. The full Joule's range is on the pumps, of course. Showplace for Spitting Feathers, the **Brewery Tap** occupies a

[22]

medieval hall that you enter via a flight of steps. Inside, it's a splendid high-ceilinged space that oozes history and, even better, good beer, with eight cask lines and eight keg pouring a generous number of guests along with the brewery's own. The **Spitting Feathers Brewery** itself, on the outskirts of the city, hosts Brewbarn Sessions, with lashings of beer, pizza and live music.
www.instagram.com/bearandbillet.
www.joulesbrewery.co.uk/our-taphouses/
  cross-keys.
https://spittingfeathers.co.uk/the-brewery-tap/.

The **Cavern of the Curious Gnome** is in the Rows above a wine bar. It's another atmospheric spot to drink beer, an imagined gnome's grotto. Fifteen keg lines are dominated by Belgian classics, including, of course, the gnome-branded La Chouffe, plus other European brews and some British craft. Four handpumps offer interesting cask alternatives. **Commonhall Street Social** is down an alleyway where up to 10 craft lines and a couple of cask give you a good range to choose from in a lively, youthful setting. Back on the main road among the shops, **Beer Heroes** is a craft beer café with a dozen lines mixing some of the best of UK-brewed craft with continentals. It hosts beer and cheese matching evenings, among other events.
https://thecavernofthecuriousgnome.co.uk/
  gnome-page.
www.commonhall.co.uk.
www.beerheroes.com/tap-room.

Brunning & Price's second pub-restaurant in the city takes a prime position looking over the racecourse with plenty of outdoor seating where you can enjoy the view. **The Architect** is so called because it's in a Regency house designed by local architect Thomas Harrison (who also designed new buildings in the grounds at Chester Castle). As for the beer, six handpumps pour cask ales including from Cheshire's Weetwood Brewery.
www.brunningandprice.co.uk/architect.

Telford's Warehouse [23] and its beer terrace occupy a grand spot where the canal broadens just north of the City Walls. Also a music and arts venue, the spacious interior is split over two floors and has well-stocked bar with six handpumps regularly pouring Weetwood and Salopian ales,

[23]

[24]

craft taps typically featuring brews from Cloudwater and Mobberley, and a decent range of bottles in the fridge. Back in the centre, the **Pied Bull** started life a thousand years ago as a coaching inn. Refurbished in 2023, it combines pub, brewery and hotel with up to five handpulls pouring a range of ales brewed on site. www.telfordswarehousechester.com/. www.piedbull.co.uk/.

👓 WHAT TO SEE & DO

Handily, there are two ancient structures here that provide a ready-made route around this city: the Roman and medieval **walls** and the **Rows**, which have developed since the Middle Ages. The walls offer a literal overview of most of the sightseeing, allowing you to pick and choose what you fancy exploring as you go – the Cathedral, Castle, Roman Garden and Amphitheatre, River Dee, canal and the racecourse (at the information point here find out why the term 'gee-gees' comes from Chester).

It's thanks to these walls that the **Grosvenor Museum** has its best exhibit, Roman tombstones that had been used as ballast in the Middle Ages and were dug out by the Victorians. The simple pictures carved on these stones are an extraordinary insight into the lives of soldiers, slaves and ordinary folk who lived here two millennia ago. A good starting point on the 3.2km (2-mile) circuit is the flamboyant Victorian **Eastgate Clock** [24], an obvious photo opportunity and, apparently, only beaten by Big Ben for horological selfies. https://grosvenormuseum. westcheshiremuseums.co.uk.

We have the Anglo-Saxon warrior Aethelflaed to thank for Chester's post-Roman existence. She pitched her fort where **Chester Castle** stands, here on the frontline with the Celts, though what you can see today dates from William the Conqueror and the Georgians. Most of it is used by the Crown Court and university; however, English Heritage frequently opens the Agricola Tower, a 12th-century gateway with a chapel decorated with medieval wall paintings. www.english-heritage.org.uk.

If wall paintings are your thing, there's a faded one from the 14th century showing St John with a book in his hand on a pillar in the fine pre-Conquest church in his name (the ruins at the church's eastern end are a hopelessly romantic

sight). This was the **Cathedral** until 1541 when that role was bestowed on the sandstone hulk in the city centre, which is built on the foundations of a shrine to St Werburgh and a Benedictine monastery. The cathedral's hit parade includes its lovely cloisters, the quire stalls and their carvings (look for the chap drinking a tankard of ale), and the modern bell tower in the grounds, a freestanding 1970s structure called the Addleshaw Tower, known locally as the Chester Rocket for its shape.
https://stjohnschester.uk.
https://chestercathedral.com.

In the cobbled heart of the city centre, you can use the medieval **Rows** as a guide. These unique, balcony-style, covered walkways at first-floor level snake through the fabric of the four main shopping streets. It's an architectural treat – medieval, Tudor, Georgian, Tudor Revival, and Brutalist, underpinned in parts by hefty original pillars of the Roman fort of Deva Victrix (see the ones under Pret a Manger on a tour with the city guides). The Rows are full of stories, one of which is how jewellers congregated where Eastgate and Bridge Streets meet – Chester had an assay office until 1962. One of the old family firms, **Lowe & Sons**, has a small permanent exhibition on the first floor of its beautiful shop that recounts the jeweller's 250-year history. Just ask, they'll let you go up there.
https://chestertours.org.uk.
www.loweandsons.co.uk.

That was then and this is now. To return to the present day, eat tacos or buy the dog some pyjamas at **Chester Market**, which has been revamped as a huge food hall surrounded by assorted shops, as is the vogue. Or catch a show or a film in **Storyhouse**, a reimagined Art Deco cinema that now houses an independent cinema, theatre, public library and café. In summer, theatre performances move outdoors to the **Grosvenor Park Open-Air Theatre** next to the river, where **boat cruises**, **pedalos**, **paddleboarding** and the **Snugbury's** ice cream parlour create a holiday atmosphere.
https://newchester.market.
www.storyhouse.com.
www.grosvenorparkopenairtheatre.co.uk.
www.chesterboat.co.uk.
www.chesterboathire.co.uk.
www.deeriverkayaking.com.
 https://snugburys.co.uk.

WHILE YOU'RE HERE

Chester Zoo, one of the country's most popular attractions and a leading global conservation centre, is on the outskirts of the city – at time of writing, it was building holiday lodges with views of the wildlife. The **National Waterways Museum Ellesmere Port** is also nearby in Thomas Telford's historic canal basin.
www.chesterzoo.org.
https://canalrivertrust.org.uk.

FURTHER INFORMATION

For more ideas on what to see and do, go to www.visitcheshire.com. Check opening times of beer venues and attractions before you go.

Winter Watch and Saturnalia parades: Time is suspended one night each December as Chester is overrun by Roman centurions and the ghoulish entourage of the Lord of Misrule, who parade and dance about its streets to the unrelenting beat of drums. Nab a place on the Rows for the best view of the commotion.
www.midsummerwatch.co.uk/winter-watch-parade.

THE TEMPERANCE INN

The beer flows freely at The Temperance Inn, Ambleside, courtesy of Inn Collection. Built in the 1800s as a base from which the Temperance Society could encourage abstinence among the locals, the large open-plan hall is a place where people now gather to drink and dine. The flint and stone façade provides the strongest visual link with the past. The 27 en-suite guest rooms are modern, a mix of singles and doubles set across three upper floors – some with fine views of surrounding fells – and there's a garden cottage. Plain in style with judicious use of patterns in the décor, the rooms are clean and comfortable with good-quality furnishings and the service is prompt and friendly. www.inncollectiongroup.com/temperance-inn.

DOGHOUSE

In the busy heart of Manchester, between Beermoth and the Northern Quarter, Brewdog opened this craft beer hotel in 2021. The 18 modern, high-tech, beer-themed rooms occupying the two upper floors unusually have their own record player and selection of vinyl – plus a guitar, in case you want to make your own music. (Fellow residents might be relieved there isn't a drum kit.) Guests receive a complimentary glass of beer on arrival – and another when you reuse your bath towel – and have access to round-the-clock hotdogs. There's a beer fridge in the shower if you can't wait to get to the bar downstairs or the roof terrace above, which has an additional 10 lines. On-site beer masterclasses are available, and yes, of course it's dog-friendly. www.brewdog.com/uk/doghouse-manchester-hotel.

MORE PUBS WITH ROOMS

KENDAL
The Shakespeare
www.shakespearekendal.co.uk.

AMBLESIDE
The Ambleside Inn [25]
www.inncollectiongroup.com/ambleside-inn.

The White Lion
www.innkeeperscollection.co.uk/hotel/
   the-white-lion-hotel-ambleside-cumbria.

GRASMERE
The Swan
www.inncollectiongroup.com/the-swan-grasmere.

Tweedies
www.tweediesgrasmere.com.

KESWICK
The Kings Arms
https://lakedistricthotels.net/kingsarms.

The Royal Oak
www.royaloakkeswick.co.uk.

Inn on the Square [26]
www.innonthesquare.co.uk.

For Lake District also see 'Stay in a Brewpub' and
'Hotel Review: Temperance Inn'.

MANCHESTER
The Oxnoble
www.stonegategroup.co.uk/hotels.

The Abel Heywood
www.abelheywood.co.uk.

Lower Turks Head
www.joseph-holt.com/pubs/lower-turks-head.

LIVERPOOL
All Bar One
www.allbarone.co.uk.

CHESTER
The Pied Bull [27]
www.piedbull.co.uk/.

The Boathouse
www.theboathousechester.co.uk.

[25]

[26]

[27]

MIDLANDS

Opposite: Birmingham, Gas Street Basin.

PEAK DISTRICT

There's a wealth of rural pubs to explore in the national park. We've chosen a route from New Mills to Matlock where you'll find a good variety of beer destinations both traditional and modern. Revolutionary mills, historic spa towns and the home of a pudding that will satisfy the sweetest tooth are also on the menu here.

NEW MILLS

 ### WHERE TO DRINK

One of the country's boldest brewers can be found at New Mills Marina. **Torrside Brewing** specialises in smoked beers and hop monsters and hosts an annual Smokefest that attracts devotees of the style from all over. Its taproom opens during the warmer months and it's worth timing your trip accordingly. Torrside also runs the town's micropub, **The Beer Shed**, on Market Street, a curious long, narrow room with a tiny bar pouring a rich variety of cask and craft beers including, but far from exclusively, its own. Its beers can also be found at the **Torr Vale Tap** within the town's historic mill.
www.torrside.co.uk.
www.facebook.com/beershednewmills.

WHILE YOU'RE HERE

Thornsett Brewery & Hop Yard [1] is a few miles out of town – literally in a field with cows. It comprises a shed with a hatch from which cask ales brewed a few feet away are dispensed, and bench tables where you can commune with the cattle if you're so disposed. Thornsett also grows its own hops on the farm. Open to visitors most summer weekends.
www.thornsettbrewery.co.uk.

 ### WHAT TO SEE & DO

Set around a gorge where the **rivers Goyt and Sett** meet, this town became a centre for the cotton industry in the 1800s. Today, the locals make Love Hearts in the former mills for sweets manufacturer Swizzels, something to suck on while you consider the ingenuity required to exploit the power of the water at the bottom of the chasm known as **The Torrs**.

Take the steep steps down from Rock Mill Lane to where the two rivers meet, passing climbers picking their way up the cliffside and beneath the towering arches of a viaduct carrying traffic almost 30m (100ft) above. Yousoon come across the atmospheric ruins of Torr Mill and Archie the screw turbine, also known as **Torrs Hydro**, which is celebrated as the first community owned and funded hydro-electric scheme in the country. From here, continue to the meadows or double back to walk upon a feat of 20th-century engineering, the **Millennium Walkway**, a path suspended over the rushing waters which offers a close-up view of the historic **Torr Vale Mill**.
www.derbyshire.gov.uk.

THORNSETT
BREWERY & HOPYARD

[2]

BUXTON

WHERE TO DRINK

This spa town is, of course, the home of **Buxton Brewery**, among the most admired of the new wave of UK craft brewers with a range of punchy hop-forward beers. It has three venues here [2]. The brewery itself, about a mile out at Staden, has a bar called **Trackside** across the yard with a grassy beer garden and fabulous views across rolling fields. Bang in the centre of town, the company's **Tap House & Cellar Bar** is two bars in one, with the widest selection of Buxton's beers on tap. www.buxtonbrewery.co.uk.

Macclesfield's **RedWillow Brewery** has an outpost in the middle of Buxton, a pub occupying a grand building that retains the air of the bank it once was, complete with a little manager's office in the corner and original etched glass and panelling, plus a balcony. Three or four of the brewery's ales are available on handpump, and there are another 18 taps on the back-bar. https://redwillowbrewery.com/pages/buxton.

Around the corner you'll discover **Beer & Bean**, a continental-style café serving good coffee, sweet waffles, sharing boards and, most importantly, a small but well-chosen range of draught beers from Belgium, Britain and elsewhere. Next door the **Frog on the Bine** is a beer café that opened summer 2024 with three cask lines and give kegs, and there's also a little bar nearby called **Seasons** which as interesting beers on tap from the likes of Manchester's Sureshot and Buxton's own Silent Brew Co. www.beerandbean.co.uk. the frogonthebine.wixsite.com. seasonsbuxton.co.uk

Buxton's micropub is **The Ale Stop**, a simple yet smart bolt-hole with the local brews you'd expect on the pumps and, unexpectedly, a DJ booth – music is as important as the beer here. And across the road there's the **Cheshire Cheese**, a comfortable Titanic Brewery house with a strong range of ales.

https://whatpub.com/pubs/HIP/660/ale-stop-buxton.

www.titanicbrewery.co.uk/our-pubs/the-cheshire-cheese.

👓 WHAT TO SEE & DO

Higher Buxton feels like a typical rural town, with a market square and shops presided over by a Victorian town hall. But walk downhill for the extraordinary townscape that still attracts thousands of visitors each day. Stand on **The Slopes**, a grassy bank landscaped in the early 1800s, to see what there is in this spa resort developed during the Georgian and Victorian eras.

Ahead stands the original complex – **Buxton Crescent**, Bath's Royal Crescent lite, flanked by the **Natural Mineral Baths** (now part of the swanky hotel that occupies the Crescent) and **Buxton Baths** (a shopping arcade, worth peeping in to see the original Minton tiles). Across the rooftops is the **Devonshire Dome** – bigger than St Paul's, they say – while westwards lie the graceful **Pavilion Gardens**, still a place to promenade, with its glass

Pavilion inspired by the Crystal Palace and Frank Matcham's later **Opera House** [4]. Taste the waters for yourself at the source, **St Ann's Well** [3], or from the fountain in the recently restored **Pump Room**.
www.buxtonmarkets.co.uk.
https://ensanahotels.com.
Buxton Baths, 4 The Square, SK17 6AZ.
Devonshire Dome, 1 Devonshire Road, SK17 6RY.
https://paviliongardens.co.uk.
https://buxtonoperahouse.org.uk.
https://buxtoncrescentexperience.com.

[4]

BAKEWELL

 WHERE TO DRINK

Thornbridge Brewery [5] is reason alone for the beer lover to visit Bakewell. Follow the river out of town (there's a footpath at the end of Bridge Street) and in 15 minutes you'll arrive at its spacious taproom. The bar showcases the Thornbridge range, including specials, on cask and keg, and there's a shop selling an even wider range of bottles and cans to take away.

Thornbridge has expanded into one of the country's leading craft brewers during the past decade, but it's held on to its original kit, which is behind the bar and serves as the focus for the **Thornbridge Experience**, a tutored tasting that delves into the details of the brewing process. Since 2024 it's also been home to one of the country's few remaining union brewing sets, and has been recreating this old brewing method. https://thornbridgebrewery.co.uk/pages/tap-room.

Back in town, the **Joiners Arms** is a good micropub with a rustic feel, pouring mostly local ales from seven handpumps and hosting live music, while the **Queens Arms** offers a decent choice in a regular pub setting and **The Manners**, on the way out east, is a solid, comfortable Robinsons' house with a full food menu, rooms and a pleasant beer garden.
www.facebook.com/JoinersArmsBakewell.
www.facebook.com/QueensArmsBakewell.
www.robinsonsbrewery.com/pubs/the-manners-bakewell.

[6]

WHILE YOU'RE HERE

The **Flying Childers Inn** at Stanton-in-the-Peak is worth a trip out. Named after an 18th-century racehorse, it's a family-owned, unspoilt, two-bar village pub that serves a great pint of Bass alongside a few other well-kept cask beers. Not to be confused with the restaurant of the same name on the Chatsworth estate.
https://flyingchilders.com.

WHAT TO SEE & DO

Who makes the most authentic Bakewell Pudding – **The Bakewell Tart Shop** or **The Old Original Bakewell Pudding Shop** (other brands are available)? Both claim to bake the best pastry filled with eggs, ground almonds and jam. And would you prefer pudding or, heaven forbid, the imposter tart? Whatever you choose, first you must earn it by taking a look around.

Amid the pretty limestone and sandstone buildings and courtyards there's medieval **Bakewell Bridge** [6] over the River Wye, and church tourism devotees will like the two Saxon crosses and 14th-century alabaster monument at **All Saints**.
www.bakewelltarts.co.uk.
www.bakewellpuddingshop.co.uk.
Bakewell Bridge, Bridge Street, DE45 1DS.
All Saints Church, South Church Street, DE45 1FD.

The bath house in **Bath Gardens** is evidence of the failed attempt to establish a spa resort to rival Buxton in the 19th century. But it's not so much the individual sights as the whole scene that makes this market town a pleasure to visit. It's a good base, too, for trips out to nearby **Chatsworth House** and **Haddon Hall**.
Bath Gardens, The Square, DE45 1BT.
www.chatsworth.org.
www.haddonhall.co.uk.

MATLOCK

WHERE TO DRINK

Handy for the station, among the shops and restaurants in Dale Road, **Twenty Ten** is a café/bar-style operation serving one of the widest ranges of draught ales and craft beers in town, rivalled only by **The Newsroom**, a former newsagents that was converted into an L-shaped micropub by an owner who realised he'd much prefer selling beer. Both also have interesting cans on the shelves. http://twentytenmatlock.co.uk. www.facebook.com/thenewsroommatlock.

To reach the Newsroom you'll need to climb halfway up Bank Road, the steep hill that runs from the bottom to the top of town, but if you need an extra excuse, **The Farmacy** is only a few doors away. Another micropub, it serves as the taphouse for **Aldwark Artisan Ales**, a few miles into the Peaks, and is one of the rare places you'll find its beers on draught. The brewery itself opens to the public only on summer weekends. www.facebook.com/thefarmacymatlock. https://aaabrewery.co.uk.

For those with the energy to scale further heights, the **Thorn Tree** awaits. It's a classic two-bar pub with great views across the Derwent Valley from the front terrace, part of Greene King but with guest beers including Bass, Timothy Taylor Landlord and local brews. www.facebook.com/TheThornTreeInn.

At the bottom of the hill, in the part of town known as Matlock Green, the **Red Lion** [7] is the home of Moot Ales, after one of the old names for the town and brewed in the shed out the back. Its range has developed well and you'll usually find four or five on cask alongside one or two interesting guests. www.theredlionmatlock.co.uk.

A mile from the centre of town, near Lumsdale Falls, **Bentley Brook Brewing** has a lovely taproom adjoining an unusual beer garden raised high above the road. Open weekends, it's usually pouring a couple of its beers on cask and keg. And a few paces away you can watch a live glass-

[8]

blower at work at **Lumsdale Glass** and browse the beautiful results in the showroom. www.bentleybrook.co.uk. www.lumsdaleglass.co.uk.

WHILE YOU'RE HERE

Off the Chesterfield road, in the village of Ashover, the **Old Poets' Corner** is a great pub, the original home of Ashover Brewery, and is currently being well looked after by Titanic. The 10 handpumps serve guests alongside the Stoke brewery's range. www.titanicbrewery.co.uk/pubs-and-bods/pubs/ the-old-poets-corner.

WHAT TO SEE & DO

Matlock may not equal spas in the modern mind, but it was a serious player in the 19th century. The son of the cotton industrialist John Smedley caught the hydropathy bug and capitalised on the local thermal waters. From his home, **Riber Castle** [8], the neo-Gothic pile you'll spy above town, John Jr commissioned **Smedley's Hydro**, on that steep road up the town, a still impressive slab of stone with a tower more suited to a French chateau. Soon a cable tramway was built to haul guests up from

the station. The trendsetting hydro is now the council offices and other existing examples, such as **Rockside Hydro**, are similarly only viewable from the outside.

Matlock Bath is a village that looks like a landlocked seaside resort with its amusement arcades, chippies and sweet shops. Dwellings terrace up the steep wooded slopes, over which flies a cable car from the hilltop country park **The Heights of Abraham**. Continue on to Richard Arkwright's **Masson Mills** (where you can pick up a quirky souvenir such as a Wound Pirn) and **Cromford Mills**. The latter – the first modern factory – is an impressively preserved complex of mill buildings and workers' housing, part of the **Derwent Valley Mills UNESCO World Heritage Site**. www.heightsofabraham.com. www.massonmills.co.uk. www.cromfordmills.org.uk. www.derwentvalleymills.org. www.johnsmedley.com/outlet.

FURTHER INFORMATION

For more ideas on what to see and do, go to https://visitpeakdistrict.com. Check opening times of beer venues and attractions before you go.

NOTTINGHAM

Nottingham is among the most diverse of Britain's beer cities, offering a seat at a couple of the country's oldest pubs alongside cutting-edge craft breweries – and plenty going on in between. Too often overlooked on the tourist trail, this city is the stuff of adventures, with tales of Robin Hood, a castle and a network of caves among the sights to explore.

CENTRAL

WHERE TO DRINK

A key component of what makes Nottingham so tempting for the beer drinker is **Castle Rock Brewery**, which stands almost within sight of the station in some old industrial buildings behind the **Vat and Fiddle** [9], its splendid brewery tap, pouring no fewer than 13 cask beers. Brewery tours are available, too. The company operates more than 20 pubs that each sell a wide choice of beers, not just its own, embracing craft keg as well as cask.

[9]

The closest is the **Canalhouse** [10], a pioneer of the country's craft beer scene, which straddles a branch of the Nottingham & Beeston Canal. Once inside you cross a bridge – and a barge – to reach the main bar and 19 lines of craft and cask. And nearby, occupying a former sweetshop, the **Barley Twist** offers 10 lines of craft and a couple of handpulled ales over two floors, while a departures screen warns that you're about to miss your train.
www.castlerockbrewery.co.uk.
www.castlerockbrewery.co.uk/pubs/vat-and-fiddle.
www.castlerockbrewery.co.uk/pubs/canalhouse.
www.castlerockbrewery.co.uk/pubs/barley-twist.

Within the station buildings, the Brew Cavern bottle shop opened its drink-in arm, **Brew Tavern**, in 2024. Five cask lines focus on modern brewers such as the local Liquid Light while seven keg taps might feature the likes of Verdant, Beak and Cloudwater plus Nottingham craft from Bang the Elephant, for instance, and US imports alongside its own lager brewed by Camerons.
https://brewcavern.co.uk/brew-tavern.

Widely acknowledged to be Britain's oldest pub, dating from 1189, **Ye Olde Trip to Jerusalem** is built into the rock on which Nottingham Castle

[11]

stands and its honeycomb of bars are really pink-walled caves. With that sort of unique attraction it's a nice surprise to find the Trip's custodians at Greene King Brewery have made an effort to provide a range of ales beyond its own. And the same is true at the comparatively youthful 15th-century **Bell Inn** [12], which has a main bar at the back, two more at the front and a wealth of architectural features among its panelled rooms. It offers a rotating choice of guest beers, including from Nottingham Brewery.
www.greeneking.co.uk/pubs/nottinghamshire/
    ye-olde-trip-to-jerusalem.
www.greeneking.co.uk/pubs/nottinghamshire/
    bell-inn.

Housed in a characterful 15th-century building, the **Angel Microbrewery** is another of the city's historic pubs that, as the name suggests, brews its own beer. Expect to find modern and experimental cask styles (branded with some curious artwork) and guests on cask and keg. Also in the Lace Market district is Castle Rock's **Kean's Head** [11]. This place understands what a beer house can be with 18 keg lines, which might include one or two rarities – a Cantillon lambic was on at last visit – and half a dozen cask pumps.
www.theangelmicrobrewery.co.uk.
www.castlerockbrewery.co.uk/pubs/keans-head.

Tucked down an alleyway among the shops, **Junkyard** and **Kilpin Beer Café** are right next door to each other, and are, in fact, owned by the same people. You're welcome to buy a beer in one and drink it in the other. Junkyard's 15 taps are focused on British-brewed craft beers from leading names such as Lost & Grounded, Polly's and Drop Project, and there are plenty more in the fridges. In a more traditional atmosphere, Kilpin brings together a range of continental beers, including German lagers and Belgian classics, with a few well-chosen cask ales. Burning Sky and Black Iris were both represented at last visit. It also hosts tutored tastings, giving customers a chance to try a selection of Trappist beers, for example.
www.junkbars.com.
www.thekilpin.co.uk.

WHAT TO SEE & DO

Arrive by train and your adventure begins in the unlikely surroundings of a multimillion-pound transformation of the old shopping mall Broadmarsh Centre, an area that must be passed through to reach the city centre. This building site is still a long way off completion at the time of writing; return visitors might like to see how it's getting along, otherwise hasten to the centre.

The taste of the 19th-century city planners for Gothic Revival and Old English styles is a little easier on the eye, illustrated by 100-plus buildings designed by the architect with a name to savour, **Watson Fothergill**. Following his trail through town is simple thanks to a free guide on Visit Nottinghamshire's website that will take you around the centre, passing through the old industrial hub of the Lace Market.

Spot the pepperpot turrets WF added to one of the warehouses that forged the city's global reputation as a centre for the delicate textile. And on his building for the *Nottingham Express*, where the author Graham Greene was a young journalist, carvings of the faces of the Liberal politicians of the day reveal the editorial line. Along the way, you'll see a statue of another famous person associated with the city, the celebrated former manager of Nottingham Forest FC, Brian Clough.
www.visit-nottinghamshire.co.uk.
Express Offices & Express Chambers, 17–25
    Upper Parliament St, NG1 2AD.
Brian Clough statue, Old Market Square, NG1 2DT.

Nottingham's big adventure is its castle. Founded by William the Conqueror as he struggled to put down the rebellious north, the fort on Castle Rock is a 17th-century ducal palace renovated in the late 19th century after being set on fire during the 1831 reform riots. They're a defiant lot round here, as you'll discover inside, where exhibits include one on the much-maligned Nottingham activists the Luddites. And you can have a virtual stick fight in the gallery dedicated to Robin Hood – so what if it's for kids. The beloved outlaw is cast in bronze, drawing his bow, just below the castle.
www.nottinghamcastle.org.uk.
Robin Hood statue, Castle Road, NG1 6AA.

The adventure continues in the network of sandstone caves that Nottingham sits on – especially curious since they were dug out by hand in medieval times. The **Old Brewhouse Yard**, around Ye Olde Trip to Jerusalem [13] at the foot of Castle Rock, was fed by a tributary of the River Trent and the cool caves behind each house were used for making ale. The yard has been revamped and a visit is now in the castle entrance fee. Want more caves? Go to the dedicated attraction **City of Caves**.
www.nationaljusticemuseum.org.uk/cityofcaves/
    visit.

The adventure takes an artistic turn at the original **Sky Mirror** by sculptor Anish Kapoor, outside the **Nottingham Playhouse**, the city's award-winning theatre. It's not the only important modern art on show, the **Nottingham Contemporary** specialises in the period. The gallery itself is worth a look as much as what's inside; set in a sandstone cliff, this RIBA award-winning structure references the city's lace industry in an intricate design embedded in its concrete façade. There are more interesting galleries around the city to hunt out with the aid of the **Nottingham Art Map**.
https://nottinghamplayhouse.co.uk.
www.nottinghamcontemporary.org.
https://nottinghamartmap.co.uk.

[12]

[13]

[14]

SNEINTON MARKET

 **WHERE TO DRINK**

A tight little circuit of good beer venues can be found east of the city centre around the trendy Sneinton Market Avenues. At its centre is the **Neon Raptor Rap Tap** [14], a light and airy functional sort of bar that shares the space with the brewery itself. A dozen keg lines pour styles from saisons to stouts. In contrast, though just as quirky in its own way, the **King William IV** – known to locals as the King Billy – is an independent freehouse that's been on this spot since 1832. Eight handpumps showcase the best of Nottingham brewing, and you should check out the roof garden.
https://neonraptorbrewingco.com.
https://new.thekingbilly.co.uk.

Two doors down, the **Partizan Tavern** offers a more stripped-back modern space to enjoy a modest selection of well-chosen cask ales and craft beers on tap mixing familiar names such as Verdant and Ashover with the rarer Ticking Clock and Bang the Elephant. The **Fox & Grapes** [15] is a Castle Rock house that has the feel of a contemporary London pub. Eight cask lines and 14 keg feature the brewery's own, plus interesting guests.
www.facebook.com/PartizanTavern.
www.castlerockbrewery.co.uk/pubs/fox-grapes.

A bit further out, on the Robin Hood Industrial Estate, you'll find **Liquid Light Brewing**. Open at weekends, the taproom is nestled among the brewing equipment and exudes a hippie-cum-goth kind of mood. A wide variety of brews are served on keg and cask.
www.liquidlightbrewco.com.

WHAT TO SEE & DO

Sneinton Market is a former fruit market that has been taken over by a new generation who have opened stores, studios and, of course, bars, in the long, low 1930s buildings. It's a place to shop for pre-loved clothes, designer pet accessories and botanical spirits, and its avenues fill with stalls and street-food carts during market days and special events.
www.sneintonmarketavenues.com.

SNEINTON MARKET
FOX & GRAPES
FOX & GRAPES
CASTLE ROCK
SNEIN
eMAR

[16]

TRENT BRIDGE

 ## WHERE TO DRINK

South of the station beside the canal the **Trent Navigation** is the tap house for Navigation Brewery. A spacious traditional pub done out in a modern style, it serves the entire available range in cask and keg, plus guests, with an impressive 13 handpumps on the bar. Castle Rock's **Embankment** [16] occupies a grand Grade II-listed building that originally opened in 1907 as a Boots pharmacy (this is the company's home town). A dozen cask beers and 10 craft keg are poured across multiple rooms.
www.trentnavigation.com.
www.castlerockbrewery.co.uk/pubs/
    embankment/.

In prime position on the riverbank with views of the Trent from its terrace, **Brewhouse & Kitchen** offers everything you'd expect from the award-winning chain in spacious surroundings that incorporate its own brewery. There's a wide choice from cask and keg, most of them created on site by a brewer encouraged to express themselves through the medium of beer.
www.brewhouseandkitchen.com/venue/
    nottingham.

WHAT TO SEE & DO

Refreshed by the beers around Trent Bridge, you'll be close to the station for your next quest, to hop on a train to Attenborough. Within 20 minutes you'll be at the gates to the **Attenborough Nature Reserve** [17], a mosaic of lakes and woodlands alive with birds. An easy walk on the circular trail will take you around the pools and islands and, depending on the time of year, you might spot some usually scarce waterfowl such as sawbills.
www.nottinghamshirewildlife.org.

[17]

CANNING CIRCUS

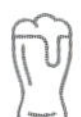

WHERE TO DRINK

Three good beer pubs gather around the gyratory system south of Nottingham General Cemetery. The **Sir John Borlase Warren** is a flagship for the local Lincoln Green Brewing and is a comfortable, welcoming multiroomed place with lots of enticing spots to sit, including a pleasant beer garden out the back. The brewer's range of cask ales plus its Blackshale craft beers are all on the bar. www.lincolngreenbrewing.co.uk/the-sir-john-borlase-warren.

The **Good Fellow George** frequently features on cask the work of Lenton Lane Brewery, on the outskirts of the city, alongside big name craft players, including local Neon Raptor, on eight keg lines. Run by Nottingham's own Blue Monkey Brewery, the **Organ Grinder** is a great place to taste the range with up to nine cask lines pouring. www.thegoodfellowgeorge.co.uk. www.bluemonkeybrewery.com/organ-grinder-pubs/nottingham.

From here, hop on a tram that stops right outside **Black Iris Brewery** [18], one of the country's cutting-edge craft producers. Its tapyard, heated in winter and girded with foliage, is open weekends when a slightly Satanic selection of hop-

[19]

packed delights (Don't Fear the NEIPA is a great beer name) are served from a shed. And almost next door you'll find the **Lion at Basford**, a champion of local brewers that has up to 10 cask ales pouring – plus a good view of the historic Shipstone's Brewery standing proudly on the hill. https://blackirisbottleshop.co.uk. www.thelionatbasford.co.uk/.

And it would be remiss not to jump off the train at Beeston, two minutes out, and open the little gate at the end of the platform that leads straight into the beer garden of the splendid **Victoria Hotel**, where you'll find 14 cask ales rotating on the pumps. https://vichotelbeeston.co.uk.

WHAT TO SEE & DO

At last, it's time to unleash your inner Robin Hood amid the trees. Not Sherwood Forest but **Nottingham Arboretum** [19], a collection of more than 800 specimens which, it's said, inspired JM Barrie's Neverland. OK, so that's not as many as **Sherwood Forest**'s 1,000 ancient oaks, but you can visit them easily from here, too. www.facebook.com/NottinghamArboretum. www.visitsherwood.co.uk.

FURTHER INFORMATION

For more ideas on what to see and do, go to www.visit-nottinghamshire.co.uk. Check opening times of beer venues and attractions before you go.

[18]

DERBY & BURTON UPON TRENT

Linked by a short train ride, Derby and Burton upon Trent share a beer scene that's mostly about traditional cask beer. Derby tells its industrial story well in an innovative local museum, and while Burton might be hiding its light when it comes to the story of brewing, it has some beautiful nature to explore.

DERBY

 ### WHERE TO DRINK

Derby has long been famous as a beer destination and two breweries in town welcome the public. **Derby Brewing** offers tours on Saturdays including lunch and a chance to chat with the brewers, while **Dancing Duck** throws open its doors for Nights in the Brewery on summer dates.
www.derbybrewing.co.uk.
www.dancingduckbrewery.com.

Arriving at the station, you don't have to walk far to find a pair of its best pubs. Housed in a wedge-shaped building with multiple rooms in shades of brown and lots of comfortable corners, the **Brunswick Inn** [20] is a brewpub that's owned by Everards Brewery of Leicester. You'll find a wide range of home-brewed beers on the pumps that you're not likely to see elsewhere, plus guests. On the next corner is the **Alexandra Hotel**, fondly known by locals as the Alex, and a long-time stalwart of good beer in the town. Nottingham's Castle Rock Brewery curates a great range including six on cask and a few craft keg options, too. A station-style digital clock aims to make sure you don't miss your train, but doesn't always succeed.
www.brunswickderby.co.uk.
www.castlerockbrewery.co.uk/pubs/alexandra-
   hotel.

Follow the Derwent for the **Smithfield Alehouse**, a smart, well-run pub on the riverbank with an unusual, curved wall and tables turned towards a long bar serving 10 cask ales, including some adventurous selections from new breweries.

[20]

[21]

Further on, the **Exeter Arms** acts as the tap house for Dancing Duck Brewery. It's a pub with bags of character and it can get very busy. Expect a variety of Dancing Duck cask ales on the bar. A few paces away, occupying a prominent, orange-bricked corner site opposite the bridge, **The Royal Standard** [21], formerly Derby Brewing's Tap House, is a grand Victorian setting for the consumption of a great range of cask beers sourced from around the region by new owner Pub People.
www.smithfieldderby.co.uk.
https://exeterarms.co.uk.
The Royal Standard, 1 Derwent Street, DE1 2ED.

For a different experience, **Suds & Soda**, by Derby Museum & Art Gallery, is a trendy craft beer bar operated by quirky Nottingham brewer Neon Raptor and showcasing on 10 lines an exciting array of IPAs, lagers and, especially, sours from the UK's leading names including The Kernel, Verdant and Vault City at last visit. There's a well-stocked cans fridge, too.
www.facebook.com/sudsandsodaderby.

Further out on Friar Gate, on the edge of the town centre, **The Greyhound** is an 18th-century pub that's retained its original oak beams and fireplace. It joined the Pub People stable as one of the company's specialist beer houses in 2024.
The Greyhound, 75–76 Friar Gate, DE1 1FN.

Heading north, the **Five Lamps** [22] is a buzzing local with almost the feel of a gentleman's club thanks to comfortable seating and design details to catch the eye, not to mention the dozen well-kept cask ales that mingle modern with traditional styles, plus six craft taps.
www.facebook.com/thefivelampspub.

Back on the riverbank, the **Furnace Inn** is a surprising find on the edge of a housing estate. The former home of Little Eaton's Shiny Brewery has a remarkable range of beers for a community boozer, not only eight cask lines, led by Shiny, but some well-chosen craft and a great can and bottle menu. All served in the friendly atmosphere you'd expect from a good local.
www.instagram.com/furnaceinn.

Derby's first micropub, **Little Chester Ale House**, was spruced up by new owners at the end of 2023 and is now a smart example of the genre. The bar, up a short flight of stairs, pours the kind of locally brewed cask ales you'd expect. www.facebook.com/LittleChesterAleHouse.

Finally, it's worth a trek south into the backstreets off Normanton Road for a look into **The Falstaff** [23], an imposing brewpub with some original Victorian features combined with quirky décor and, of course, cask ales brewed on the premises. Don't forget to check out the historic adverts on the way to the toilets. www.facebook.com/TheFalstaffDerby.

👓 WHAT TO SEE & DO

For three centuries, Derby has been all about engineering, and even before you switch off your engine, this city's fascination with technology becomes apparent. If you visit Derby by car, you might want to leave it in the 'Safest Car Park in the World'. This confident claim is made by the aptly named Bold Lane Parksafe, because not only is it staffed 24 hours a day, but each parking bay has CCTV.

Bold Lane Parksafe, 13 Bold Ln, Derby DE1 3NT.

Derby is also the home of Lombe's Mill, which captured water from the River Derwent to power silk throwing in the 18th century. The mill is now the **Museum of Making** [24], an attraction that makes clear its intent to entertain as well as educate the moment you step through the door – hanging above your head is a deconstructed Toyota Corolla and a Rolls-Royce jet engine (both companies have factories in the area and Rolls-Royce has an exhibition open to the public). Thousands of objects here reveal the materials and processes used in manufacturing and how they shaped the fortunes of the city and its people. Don't miss the photographic tableaux by the artist Red Saunders, one of which remembers the silk workers involved in The Derby Lockout of 1833–34, Britain's first major industrial dispute. https://derbymuseums.org/museum-of-making. www.rolls-royce.com/about/heritage-trust/visit/derby-and-hucknall-branch.aspx.

[22]

[23]

[24]

The museum is part of **Derwent Valley Mills**, which also includes Jedediah Strutt's complex at Belper, Richard Arkwright's Cromford Mills (see Matlock, page 82), and the workers' housing at Darley Village. It's an industrial landscape of such importance that it was awarded UNESCO World Heritage status in 2001 and its trail follows the river upstream for 15 miles.
www.derwentvalleymills.org.

But don't stride out along the riverbank quite yet. There's still the world's biggest collection of works by the 18th-century artist Joseph Wright to see at **Derby Museum and Art Gallery**. Or at least some of them, including his famous painting *A Philosopher Lecturing on the Orrery*, which shows both his deep interest in science and industry and skilful interplay of light and dark on canvas.
https://derbymuseums.org/museum-and-art-
  gallery.

Stroll along the historic streets of Sadler Gate and Iron Gate and you'll find a quirky take on tech, the **Derby Computer Museum**. Founder Rob Watson has amassed vintage computers from the past 40 years, which are plugged in ready for a game of early Nintendo and other retro amusements. Derby has form when it comes to video games. This is where **Lara Croft** was created – there's even a road named after her.
www.derbycomputermuseum.co.uk.

Iron Gate opens onto **Market Place**, behind which stands Derby's Victorian **Market Hall**, topped by an iron and glass barrel-vaulted roof designed by the Derbyshire-born architect Rowland Mason Ordish, who also designed the roof at St Pancras Station in London. At the time of writing, it was undergoing a multimillion-pound redevelopment, but you should be able to go inside by the time you read this.
www.derby.gov.uk/business/derby-market-hall.

BURTON UPON TRENT

 WHERE TO DRINK

Burton has a special place in the history of British, and indeed world brewing. Here, in 1822, the first commercially brewed India Pale Ale was produced by Samuel Allsopp, and it so successfully slaked the thirsts of the Raj that dozens more breweries sprung up to take advantage of local water that is best suited to pale beers. Sadly, Burton's brewing museum has been mothballed, but evidence of the heritage is all around you. This is still a brewing town, dominated by Molson Coors and Carlsberg Marstons, and its pubs continue to celebrate traditional beers, notably Bass. Dedicated followers of this distinctive, now scarce, ale will journey to Burton for the best examples.

Their first stop, next to the station, is traditionally the **Roebuck Inn**, a comfortable, bustling local that knows how to keep its ales, pulled from six handpumps. Close enough to be in the same postcode is the **Devonshire Arms**, larger with a more extensive range that usually includes a brew from Burton Bridge Brewery alongside an excellent pint of Bass.
www.facebook.com/roebuckinn.burton.
www.facebook.com/devonshirearmsburton.

Arguably the most sought-after Bass experience, though, is found at the **Coopers Tavern** [26] where it's poured straight from the cask. There are a few snug drinking areas here, including a bar at the front that's a shrine to Bass – even though the pub is now run by Joule's Brewery. The best place to sit, though, is on the stage right at the back where you can watch the beers flowing behind the tiny bar.
www.joulesbrewery.co.uk/our-taphouses/
   coopers-tavern/

Another special place to drink Bass is the Constitutional Club, better known as the **Cons Club**, where it's the only draught ale available. It's a membership club but it frequently opens its doors to visitors. Once you're in, turn through the door on the right into the billiard room, where you enjoy your pint in grand surroundings while watching the snooker.
www.theconsclub.co.uk.

If you're looking for more than Bass, on the same road, the **Dog Inn**, a Black Country Ales pub, has no fewer than 14 handpumps, and **Beeropolis** is a micropub probably unique in the town for leading on craft keg with a dozen lines pouring modern IPAs and other styles including sours. A couple of cask ales are usually on, too.
www.blackcountryales.co.uk/pubs/the-dog-inn.
www.facebook.com/BeeropolisBurton.

Burton Bridge Inn [25] is home to the town's smallest brewery, which is out the back (along with a skittle alley). As well as a variety of their own recipes covering all the traditional styles, it offers a take on the famed Draught Burton Ale originally brewed by Ind Coope. A new management team took over in 2024, promising to make it an even better beer destination.
www.burtonbridgebrewery.co.uk.

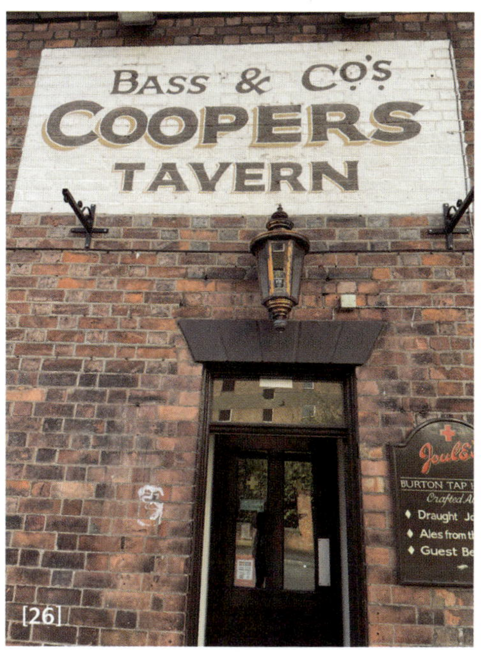

[26]

Burton, once responsible for brewing a quarter of the beer sold in Britain, really deserves more of the tourism spotlight. But you'll need to hunt out the story of beer for yourself. The National Brewery Centre closed in 2022 and while Bass Town House, the building where William Bass established his business in the 1700s, still stands, the 30-plus breweries, dating from the 1700s to the 1960s, are largely gone. You'll need a tutored eye to spot the remaining brewery buildings or the patience to plot a path using the Burton map on Historic England's website.

And you'll have to scout about for the scattering of non-beer heritage, too. There are remnants of the medieval abbey, around which the town developed, in the grounds of the 18th-century church of **St Modwen's**, and signs of Victorian boom times in the grand edifice of the neo-Gothic town hall on King Edward Place. You can see the boilers being stoked at the 19th-century pumping station at **Claymills**, and the abbots' hunting lodge, **Sinai Park House**, opens for special events.
www.theburtonthree.com/st-modwens-church.
https://sinaiparkhouse.co.uk.
https://claymills.org.uk.

Conversely, the most rewarding experience in this former hive of industry is Burton's natural surroundings. The town is bounded by the **National Forest**, and the River Trent and Trent & Mersey Canal run through it. But just steps from the high street is the **Trent Washlands**, a floodplain that lightens the mood with its hay meadows and woodland where heron fish and marsh marigolds bloom. There are well-made paths to follow and you can cross the bridge over the wetlands to **Stapenhill Gardens**. There are a few sculptures to see as you go, including *Monumite*, a Marmite jar carved in Portland stone. The spread, a by-product of brewing, is made here, which, like Burton, divides opinion.
www.nationalforest.org.

FURTHER INFORMATION

For more ideas on what to see and do, go to www.visitderby.co.uk and https://discover eaststaffordshire.com. Check opening times of beer venues and attractions before you go.

BIRMINGHAM

Birmingham is arguably one of the UK's most underrated beer destinations. It's got the taprooms, the bars, the classic pubs and two or three venues that are genuinely unique, all within walking distance of the city centre. The canal hub and Jewellery Quarter are hotspots to visit but, in recent years, the city has also become an architectural showcase.

CITY CENTRE

 ### WHERE TO DRINK

Resist the urge to follow the crowds flooding out of New Street Station to the shops; instead, nip round the back to the **Craven Arms** [27]. On a quiet corner, the building stands out thanks to the deep green and gold tiling, signature of Holder's Brewery which closed more than 100 years ago. The pub is now in the safe hands of Black Country Ales, with 11 handpumps pouring a good range on cask alongside Black Country's own and a few modern pales, too. www.blackcountryales.co.uk/pubs/the-craven-arms.

Opposite each other on Waterloo Street are two brewery tap houses. **Purecraft Bar & Kitchen** is the city showroom for Warwickshire's Purity Brewing, whose own beers mingle with guests across eight cask lines and 14 keg covering all the styles. A dedicated tasting room hosts events for up to a dozen people. While much smaller, **Sommar Birmingham** still offers 20 craft taps pouring a great range of modern beers alongside some vibrant cocktails. And just around the corner you can sup amid the mirrors and chandeliers at

Thornbridge Brewery house **The Colmore**, a grandly refurbished bank with a 9m (30ft)-long bar counter serving beers from the Bakewell company and some privileged guests, six on cask and 16 from the keg taps. www.purecraftbars.com. https://sommar.co.uk/locations/birmingham. www.colmoretap.co.uk.

The **Post Office Vaults**, as the name suggests, is a subterranean bar beneath the Post Office building. It has two entrances, on New Street and, the more obvious one, on Pinfield Street, both with stairs leading down to a narrow room where those in the know can sample a spectacular range of bottled, mostly Belgian ales. There are always half a dozen cask ales on, too, in case you fancy something wetter. www.postofficevaults.co.uk.

Another Birmingham legend, the **Wellington** has long championed cask beer. Now part of the Black Country Ales group, it has 16 handpumps pulling mainly lower gravity brews from the Midlands and

CRAVEN ARMS
HOLDERS
H
ALES & STOUT
CRAVEN ARMS
HOLDERS ALES
CRAVEN ARMS
BAR

[28]

beyond. Make sure you check out the upstairs area and its roof terrace. The city's **Head of Steam** branch is not too far away, handy if there's a game on telly you want to watch but not drop the beer quality. Choose from seven cask lines and 20 keg taps that mix British craft beer with continental classics.
www.thewellingtonrealale.co.uk.
www.theheadofsteam.co.uk/bars/birmingham.

Pinball, anyone? Heading towards Moor Street Station, you'll find a wedge-shaped bar called **Tilt** in the City Arcade. At the thin end there are 18 lines of craft beer from all over the UK and the USA, too, and at the thick end, and on two more floors, you'll find pinball machines rattling and dinging away. It's unusual. And further on, under Moor Street's railway arches, **Kilder** is a chilled beer café with a rotating selection from across Britain and beyond on 14 lines. There are occasional tasting sessions and across the beer garden a sibling company serves burgers.
https://tiltbrum.com.
www.kilderbar.co.uk.

Under the arches on the railway line out of Snow Hill Station the **Indian Brewery** [29] tap house is a bright and colourful surprise, lit by neon signs and retro advertising. On the bar eight taps serve its own range of beers, brewed in Aston, including pales and IPAs alongside its flagship Birmingham Lager, all designed to accompany a food menu of curries and loaded naans. Cross over to the other side of the railway line and you will find the Attic Brewery's city centre showcase the **Barrel Store** [28], as well as the place where you watch its oak-aged brews reach maturity. Up to 16 taps pour beers brewed at Attic headquarters in Bournville, plus guests.
www.indianbrewery.com/snowhill.
https://atticbrewco.com/pages/the-barrel-store.

The Wolf completes a trio of venues close to the station. This rustic, family-owned, dedicated craft beer bar exudes enthusiasm for modern brews with some of the nation's best producers pouring from its dozen or so lines – and there's more in the well-stocked fridges.
www.thewolfbirmingham.co.uk.

ਦਲਜੀਤ ਕੈਫ

Two Towers Brewery is based at the **Gunmakers Arms**, a quirky neighbourhood pub where you can quaff its cask ales and enjoy live music and other events. Brewery tours are available and can include a Chinese meal. Around the corner, **The Bull** dates from around 1800 and has remarkably survived all the development surrounding it with its charm and character intact. There are usually three or four local cask ales on the bar plus a craft keg option and a decent choice in cans.
https://twotowersbrewery.co.uk.
http://www.thebull-pricestreet.com

South of Snow Hill Station, in the splendid Great Western Arcade, the **Good Intent** occupies a former Victorian shoe shop with gracefully curved windows and extends into a gallery area upstairs. It's a marvellous space and the beer is pretty good, too. It's owned by Craddock's Brewery in Stourbridge so naturally its own ales feature on the handpumps, and there are guests, too, on cask and keg. What's more, it's not-for-profit so you'll be donating to local charities with every beer you buy.
www.craddocksbrewery.com/portfolio/
   the-good-intent.

👀 WHAT TO SEE & DO

Exit **New Street Station** and you haven't arrived in Birmingham you've landed, because it appears you got here in some sort of spaceship if the great ribbon of shining steel wrapped around the terminus is anything to go by. Walk about the city centre and similarly audacious sights loom into view, including the award-winning **Selfridges Building** [30]. The oozing blob studded with aluminium discs, as trendsetting as the 1960s' metal dress by Paco Rabanne that inspired its design, opened in Birmingham's beating retail heart, the **Bullring**, in 2003.
www.selfridges.com.

Another vision for the 21st century is the **Library of Birmingham** (which hosts exhibitions as well as books), a stack of glass and gold boxes wrapped in a veil of circles on **Centenary Square**. Squares are something Birmingham excels in. Centenary Square is the biggest of them and not only displays the library but the **Symphony Hall** (which has a small art gallery within the same complex),

and **The Rep**, the city's main theatre.
https://birmingham.spydus.co.uk.
https://cbso.co.uk.
 www.castlefineart.com/uk/galleries/
   birmingham-icc.
www.birmingham-rep.co.uk.

Cross the bridge from Centenary to **Chamberlain Square**, recently remodelled to show off two office blocks and the spruced-up Chamberlain Memorial, as well as the **Birmingham Museum & Art Gallery**, which has one of the world's largest collections of Pre-Raphaelite art. Chamberlain leads to **Victoria Square**, site of the city council's headquarters and Dhruva Mistry's **The River**, a cascade that spills 3,000 gallons of water per minute down the centre of the square's wide staircase, surmounted by a reclining female figure nicknamed 'The Floozie in the Jacuzzi'.
www.birminghammuseums.org.uk.

There are lots more sculptures to see, from Sir Antony Gormley's Iron: Man [31] (Victoria Square) and William Bloye's Boulton, Watt and Murdoch (Centenary Square), aka 'The Golden Boys', three gold-painted bronzes of the inventors William Murdoch and James Watt and the industrialist Matthew Boulton who worked together in the 18th century. (It's a shame Luke Perry's *Forward Together*, a monument celebrating the city's multicultural community, was only temporarily on show in the centre.) Birmingham's heritage is

Birmingham Mela: Sounds and spices mingle at the city's South Asian music festival, the biggest in Europe. As well as live music there is a food area serving specialities from the region and an arts village where you can learn a few dance moves.
www.birminghammela.com.

[30]

ever-present, not just in statuary and monuments but in the Georgian and Victorian buildings that jostle for space with their modern counterparts, including, in China Town, the only remaining courtyard of the cramped **back-to-back houses** that ordinary working Brummies lived in from the mid-19th century to the 1970s. www.nationaltrust.org.uk/visit/birmingham-west-midlands/birmingham-back-to-backs.

It's a short walk to Digbeth, where street art flourishes and old industrial premises are being transformed into studio and event spaces, especially around the old Bird's Custard Factory. www.digbeth.com.

[31]

[32]

GAS STREET BASIN

WHERE TO DRINK

Take a walk west along the canal to the **Sommar Brewery & Taproom** for up to 20 lines pouring beer brewed on the spot and more – right next door to the Legoland Discovery Centre. Or for a well-kept pint of cask ale in a simple, comfortable traditional pub, the **Prince of Wales** across the water has a dozen to choose from.
https://sommar.co.uk/pub-quiz-birmingham/
    brewery.
www.blackcountryales.co.uk/pubs/the-prince-
    of-wales.

WHAT TO SEE & DO

Birmingham was an important market town from medieval times, a legacy that lives on at the Indoor, Open and Rag markets at the Bullring. But it's better known today for its extensive canal network, which turned the city into a powerhouse of industry and commerce in the 18th century. **Gas Street Basin** is the hub of these waterways, which

once thrummed with activity but is now a quiet place to admire colourful narrowboats, pick up the towpath or get on the water by boat, kayak or paddleboard with the help of **Roundhouse Birmingham**.
https://canalrivertrust.org.uk.
https://roundhousebirmingham.org.uk.

Bars, restaurants and shops have established themselves in old and new buildings on the banks of the canal here, at **Brindleyplace**, named after James Brindley, the architect of the modern canal system, and **The Mailbox**, the former Royal Mail sorting office (where, *Archers* fans, the oldest of the soaps is recorded). Amid them, in a neo-Gothic Victorian school, is **Ikon**, a gallery that provided contemporary artists with a much-needed home when it was founded in 1964 and has grown in influence ever since.
www.brindleyplace.com.
https://mailboxlife.com.
www.ikon-gallery.org.

JEWELLERY QUARTER

WHERE TO DRINK

In what was originally a 19th-century jewellery workshop, **1,000 Trades** [33] is an independent craft beer bar and community hub over three floors. Seven keg lines feature the work of brewers such as Deya, Fixed Wheel and Vault City and up to four cask lines lean towards modern styles. https://1000trades.org.uk.

Easily missed (but don't), **Rock and Roll Brewhouse** is behind an unassuming yellow street door where a giant cardboard cut-out of Elvis Presley points you up the stairs to a bar buzzing with life. Three cask ales brewed on the floor below are poured from pumps adorned with a miniature record, there is music memorabilia everywhere you look and a vinyl playlist is themed on the day's events. www.facebook.com/RocknRollBrewBar.

Burning Soul Brewery [32] opens its modest taproom at weekends with a bar built from bottles offering 10 lines of its own craft beer, which included various IPAs, a lager and a couple of stouts at last visit. And a rather good mild was on the single handpump. Brewery tours available. https://burningsoulbrewshop.co.uk.

And the Quarter welcomed a new **Indian Brewery** taproom in the summer of 2024.

[33]

WHAT TO SEE & DO

Birmingham's Jewellery Quarter is still focused on turning out almost half of the UK's jewellery, so you're good to pick up a diamond in the shops here. You can book a tour of **JW Evans**, a former silver factory, and it's possible to go inside the School of Jewellery to see the exhibitions by contemporary designers and makers in its **Vittoria Street Gallery**. There's also contemporary art and design on display at the **RBSA Gallery** (the Royal Birmingham Society of Artists). www.birminghammuseums.org.uk. www.english-heritage.org.uk. www.bcu.ac.uk. https://rbsa.org.uk.

The Pen Museum is this quarter's real gem, charting the development of the steel nib from its Victorian heyday to the rise of the Biro. You can poke about the displays of pen paraphernalia, get to see how a nib is made, have a go at calligraphy and even analyse what dark secrets your handwriting reveals about your character. As if that museum were not obscure enough, it's topped by the **Coffin Works** around the corner, where Newman Brothers manufactured funereal furniture. For the full shadowy feel of this dark subject book a place on one of its candlelit tours. https://penmuseum.org.uk. www.coffinworks.org.

WHILE YOU'RE HERE

Sparkbrook's Balti Triangle is the place to try the eponymous curry invented in this city. Follow the heritage trail around the garden village of Bournville, built by Quaker family the Cadburys for their workers. Birmingham is rightly proud of the huge cultural contribution made by the late poet and campaigner Benjamin Zephaniah. He's now celebrated in an official mural in Handsworth Park. https://balti-birmingham.co.uk. www.cadbury.co.uk.

FURTHER INFORMATION

For more ideas on what to see and do, go to https://visitbirmingham.com. Check opening times of beer venues and attractions before you go.

WHERE TO STAY: MIDLANDS

OLD HALL INN

Daniel Capper has made the hamlet of Whitehough in the Peak District, where he grew up, a great place to drink, dine and stay. Having taken over the Old Hall Inn in the early 2000s, he expanded the premises into the house at the rear, his childhood home, opening seven guest bedrooms upstairs. Next, he bought the pub across the road, the Paper Mill Inn, extending those premises too, and adding a further eight en-suite bedrooms. Plus, he's turned a couple of nearby cottages that his young family has outgrown into self-catering options and created a studio apartment for two. So there's plenty of choice across a mix of the spaces, which vary in shape and style but are unified by comfy beds made with good-quality linen and spotless bathrooms. In the pubs, local brews are preferred at the pumps, and you've the choice of a sit-down menu in Old Hall or street food pop-ups at the Paper Mill.
www.old-hall-inn.co.uk.

THE DRAGON

Perfectly positioned for visiting both Derby and Burton, The Dragon sits in a village on the banks of the Trent & Mersey Canal, a bucolic aspect you can enjoy in the beer garden while sinking a glass of the pub's own brew, Boot Beer, made in the microbrewery at sister property the Boot Inn in nearby Repton. In recent years, Bespoke Inns has expanded the pub into neighbouring buildings to make more space for the bar and a restaurant, and there's the novel option of eating in transparent domes in the garden. Seven en-suite double bedrooms, some of which can convert to twins, occupy an adjacent cottage. The look is less village pub, more boutique hotel, the rooms having been stylishly decorated in neutral shades with pops of colour and patterns in cushions, bedspreads and drapes.
https://thedragonatwillington.co.uk.

MORE PLACES TO STAY

PEAK DISTRICT
Rutland Arms Hotel, Bakewell [34]
https://rutlandarmsbakewell.co.uk.

The Red Lion, Bakewell
www.redlionbakewell.com.

The Castle, Bakewell [35]
www.greenekinginns.co.uk/hotels/derbyshire/
   castle.

The Red Lion, Matlock
www.theredlionmatlock.co.uk.

NOTTINGHAM
Lace Market Hotel [36]
www.lacemarkethotel.co.uk.

BIRMINGHAM
The High Field [37]
www.highfieldedgbaston.co.uk.

[34]

[35]

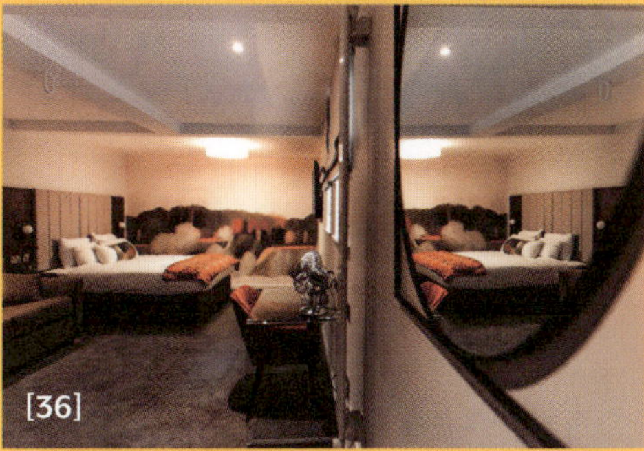

[36]

[37]

LONDON & EAST

Opposite: Crate Brewery, Hackney Wick.

NORWICH

The fact that Norwich hosts an annual month-long City of Ale Festival tells you how proud it is of its beer culture. And rightly so. It's home to some great traditional pubs and has embraced craft, too – there's great variety here. You'll also uncover many stories that will guide your way from Tombland to a medieval hall now nurturing new writing talent.

CITY CENTRE

 ## WHERE TO DRINK

Not far from Norwich Castle, the **Murderers** (nobody calls it by its proper name, the Gardeners Arms) performs the rare trick of being both a sports bar and an ale house, winning awards for both. Choose from around a dozen cask beers and some craft keg options while you're watching the big game.
www.themurderers.co.uk.

Sir Toby's Beers is a stall on the edge of Norwich Market, opposite the Guildhall. It usually has three or four craft beers on tap to drink on the spot in the open air, alongside more than 100 bottles and cans. It also hosts occasional tastings.
https://sirtobysbeers.co.uk.

If Belgian beer and food is your thing, head for the **Belgian Monk**, an atmospheric venue on Pottergate. Its extensive beer list includes 10 draught lines pouring Delerium Tremens, DeKoninck, Wittekerke, Ter Dolen Kriek and various Petrus styles. One-third flights are available and there's also a big range of bottles, of course.
www.thebelgianmonk.com.

St Andrews Brew House is a brewpub in the modern style, a large space split over two floors. Its own range of beers are available from cask and keg, and the six handpulls and eight taps also include a well-chosen guest or two. A set food menu can be paired with a flight of house beers, and brewery tours are also available.
www.standrewsbrewhouse.com.

Boasting a history dating back to the 14th century when the original building on the site was largely destroyed in The Great Fire in 1507, there are two good reasons to visit the **Ribs of Beef**: the range of beers, numbering nine cask ales and three craft brews, and the riverside terrace that looks out over the Wensum, which is a fine spot to drink them – if you can find a seat. Going west out of the city centre, **the Plough** is Norfolk's Grain Brewery's tap house in town, comprising two rooms with a bar between pouring Grain beers from half a dozen handpumps plus more from the keg taps.
https://ribsofbeef.co.uk.
www.grainbrewery.co.uk/grain-pubs.

👓 WHAT TO SEE & DO

Norwich is a city of stories (recognised by UNESCO as a City of Literature), and one of its most thrilling is Tombland, known to many more now thanks to author CJ Ransom's epic historical whodunnit of the same name. The Norse word for 'open space', Tombland was the pre-Conquest centre of Norwich town life, a good place to begin exploring. Start on Tombland Alley, a path along the old flint walls of the church of **St George Tombland** and its burial ground (they say it's so high because of the number of bodies that lie beneath from waves of plague), leading to the lopsided timber-framed **Augustine Steward House**, where the Earl of Warwick plotted to put down Kett's Rebellion, the local revolt against enclosures, in the mid-1500s. www.achurchnearyou.com. Augustine Steward House, 14 Tombland, NR3 1HF.

Beyond is the site of the Saxon marketplace and the **Erpingham Gate**, a flint-studded stone arch that locals were forced to build to keep themselves out of the monastery grounds following bitter battles with the monks. Before you go through it to the cathedral, you'll pass a memorial to the locally-born nurse **Edith Cavell**, shot while helping Allied prisoners escape in the First World War – she is buried here. This huge place of worship is lavish, with gigantic flying buttresses, an embarrassment of medieval roof bosses, fabulous cloisters (where Shakespeare plays are performed in summer), and a sky-high spire that is home to a pair of breeding peregrine falcons in the spring. The Hawk and Owl Trust has set up a nesting platform and you can watch via the webcams or from the observation point in The Close. https://cathedral.org.uk.

The battle for the marketplace was lost when the Normans moved the **market** to where it stands today and remains a landmark, a sea of colourful canopies covering 200 market stalls, which are open for business six days a week. Go around the corner to find a quite different type of market, **The South Asia Collection** [2], in the city's Victorian roller-skating rink. This bazaar shows and sells sculptures, textiles, furniture, clothes, art and crafts collected from the Swat Valley in Pakistan by owners Jeannie and Philip Millward since their travels to the region in the 1970s. www.norwich.gov.uk/norwichmarket. https://thesouthasiacollection.co.uk.

There are more stories to this city. At **Elm Hill** [1], you can stroll along Norwich's prettiest medieval street, all cobbles and crooked timber beams, and see where the Pastons lived, the family whose collection of correspondence provides us with a rare window on the early Tudor world. **The Julian Shrine** stands on the spot where Julian of Norwich wrote *The Revelations of Divine Love* at the turn of the 15th century, the earliest known book in the English language written by a woman. And Norwich is fond of the tale of that most English of Turkish-born soldiers, St George. The warrior slays the brute in a spectacular 3.7m (12ft)-high, 15th-century wall painting in **St Gregory's**, a place of worship

turned antiques emporium. A painted head and snapping jaw of the local version, Snap – a regular sight in local pageants down the years – is on display in the **Museum of Norwich at The Bridewell**. Snap's likeness is also one of the city's best examples of **street art** on the City of Stories trail. Malca Schotten's giant red beast is curled up on a wall above the corner of Red Lion Street and Westlegate [3]. **The Norwich Society** is a good source for self-guided trails that tell further stories, such as the history of the city's walls, and the museum in **Norwich Castle** [4] has more local detail. At time of writing, the Castle was undergoing a major redesign, recreating the stone keep's original Norman layout and reinstating medieval floors and rooms, including the great hall and King's chamber. It will also open a new gallery displaying more than 1,000 medieval treasures and a new lift from the ground floor to the battlements will make it the most accessible castle in Britain.
Elm Hill, NR3 1HN.
   www.julianshrine.org.
St Gregory's, St Benedicts Street, NR2 1ER.
www.museums.norfolk.gov.uk/museum-of-norwich.
www.museums.norfolk.gov.uk/norwich-castle.
www.thenorwichsociety.org.uk.

Turn the pages of Colonel Unthank's Norwich, a book based on Clive Lloyd's entertaining blog of the same name (pick up a copy at **The Book**

[3]

Hive), and read the story about the connection between some of Norwich's finest examples of Norfolk flint (see the **Museum of Norwich at the Bridewell**, which is a prime example) and the 'Public Sculpture' on **Prospect House**. Its Modernist sculptor Bernard Meadows, an associate of Henry Moore, was fascinated by how flint is composed from Norfolk's chalk bedrock and the shapes and holes in its form. His abstract work at the entrance of this Brutalist building is a good example, with its golden spheres oozing out of solid concrete blocks.
https://colonelunthanksnorwich.com.
www.thebookhive.co.uk.
Prospect House, Rouen Rd, NR1 1RE.

[4]

[5]

OVER THE WATER

WHERE TO DRINK

The **King's Head** [5] is an unpretentious pub simply serving good beer, which will typically include eight or 10 cask ales, mostly unusual selections brewed in Norfolk, and a range of Belgians. The main bar area is out the back, but the smaller front room is the best place to enjoy a pint. www.kingsheadnorwich.com.

Two backstreet locals are worth seeking out. The **Plasterers Arms** is a modest corner house that cares about its beer, always serving a broad choice of pales and darks on cask and keg plus a sour or two, and it has an extensive range in the fridge, including sharing bottles. Around the corner, **The Leopard** is a more modern pub that also mixes cask and craft and makes some very interesting selections from brewers around the country, with which you can wash down your homemade pie and mash. https://theplasterersarms.co.uk. www.theleopardpub.co.uk.

The Artichoke is a Norwich landmark, a striking, unusual piece of architecture that houses a welcoming pub with a draught beer list featuring a dozen craft brews from around Britain and Europe, plus a few more on cask. The Arti bottle shop next door has a couple of taps of its own and outside tables. And just across the junction there's a micropub, the **Malt & Mardle**, which has four craft keg lines and three cask with some interesting choices, and **Venus Vinyl**, where you can browse rare records while sipping on a canned craft beer. www.facebook.com/artichokepub. https://maltandmardle.co.uk. https://venusvinyl.com.

Further north, the **Duke of Wellington** [6] is a substantial traditional pub with a remarkably wide range of cask beers made possible by supplementing the handpumps with ales poured straight from the barrel in a glass-fronted cold room behind the bar. The owners also run the

	Brewery	Name	ABV	Tasting Notes		Price
	Wolf - Besthorpe - Norfolk	Golden Jackal	3.7	Hoppy, Thirst Quenching Golden Session Bitter	L	3.70
	Wolf - Besthorpe - Norfolk	Wolf in Sheeps Clothing	3.7	Traditional Norfolk Brown Mild	D	3.70
	Wolf - Besthorpe - Norfolk	Edith Cavell	3.7	Golden & Hoppy With Just a Hint of Blackcurrant	L	3.70
P	Wolf - Besthorpe - Norfolk	Wolf Ale	3.9	Copper Coloured Best Bitter	M	3.70
3	Wolf - Besthorpe - Norfolk	Woild Moild	4.8	Rich and Fruity Traditional Mild	D	3.90
6	Boudicca - Norwich - Norfolk	Queen Of Hops VE	3.7	Refreshing, Fruity, Rounded Pale Ale	L	3.90
	Mr. Winter - Norwich - Norfolk		3.8		L	
	Hop Bak - Salisbury - Wilts		5.0	The Original Summer Ale Brewed All Year Round	L	
	Oakham - Peterborough - Cambs.	JHB	3.8	Hoppy, Light, Crisp & Refreshing	L	4.30
	Oakham - Peterborough - Cambs.	Bishop's Farewell	4.6	Hoppy With Tropical Fruit Flavours	L	4.30
	Oakham - Peterborough - Cambs.	Citra	4.2	Fresh, Fruity & Hoppy	L	4.30
31	Nene Valley - Oundle - Northants	Manhattan Project G/F	4.0	Triple Hopped IPA G/F	L	4.30
25	Caledonian - Scotland	Deuchars	3.8	Golden Fruity I.P.A	L	3.30
9	Salopian	Golden Thread L	5.0	Golden Ale - Infusion of aroma hops L		4-50
12	Crouch Vale	Yakima Gold L	4.2	Very Pale aromatic Ameritu hops	L	4-50

Wolf Brewery in Attleborough, so expect a few of its beers to feature. www.dukeofwellingtonnorwich.co.uk.

The **Brewery Tap** is, to be specific, the Fat Cat Brewery tap house, a big old barn of a pub with no fewer than 15 cask lines and 18 keg, not only showcasing Fat Cat's beers but welcoming plenty of guests in all styles, making sure there's something for everyone. And the helpful staff know their stuff. http://fatcattap.co.uk.

WHAT TO SEE & DO

Before you cross the River Wensum make a diversion to the former premises of **Bullards' Anchor Brewery**. Its fermentation hall is now a block of flats, identified by the buffed-up lettering on its side which spells out the brewery's name. Up some steps in the car park behind the building is St Lawrence's Well, a water source granted to a brewer, Robert Gibson, in the late 1500s, as long as he provided a conduit for public use. He was so taken with his own generosity that he inscribed the stone surround with a verse singing his praises. Coslany Street, NR3 3XP.

Over the river pick up the trail of the free-to-download self-guided walks 'Norwich's Nooks & Crannies' by Norwich District Council. They cover 8km (5 miles) of historic alleys, courts and lanes and two of the three routes venture to less-trodden paths Over The Water, regaling tales of a former landscape of mills, warehouses and factories. Time your tour to allow for a diversion to **Junkyard Market** [7], where you can refuel on a choice of street food from around the globe. www.norwich.gov.uk/citywalks. www.junkyardmarket.co.uk.

AROUND RIVERSIDE

 WHERE TO DRINK

On the road heading east from the station, away from the city centre, you'll find the **Coach and Horses** (not to be confused with the pub of the same name on Bethel Street), home of **Chalk Hill Brewery**, which is in a shed at the back. At the front there is a large terrace, and inside there's a traditional ale house feel. The Chalk Hill range is joined by a couple of guests on cask. Brewery tours are available on request.
www.thecoachthorperoad.co.uk.

Keep going and you'll come to the **Fat Cat & Canary** [8], sibling pub to the Fat Cat. It's an open-plan traditional pub with a dozen pumps showcasing Fat Cat Brewery cask ales among others, and world beers on the taps.
www.facebook.com/fatcatcanary.

To the other side of the Carrow Road football stadium, the **Rose Inn** is a bright and modern pub with five cask lines and five keg pouring some interesting pales and IPAs, while the fridges offer a wider selection is all the styles, including sours and an especially good choice of alcohol-free beers. There's a delicatessen on site, too.
www.therosepubanddeli.co.uk.

Follow the road south, cross the bridge over the railway line and smartly double-back on yourself, and you'll find **Redwell Brewing** and its spacious taproom where you can enjoy a lager or a variety of pale ales brewed just feet from your seat. It also has a pizza kitchen. Brewery tours are available.
www.redwellbrewing.com.

Dragon
Hall

WHILE YOU'RE HERE

It's a bit of a stroll out along the Dereham Road, but it would be silly not to visit Norwich without calling in at the legendary **Fat Cat**, one of the nation's great beer pubs, full of character and splendid brews. Choose from 20 handpulled cask ales, which include Fat Cat's own and a mix of famous names and more obscure brews from around the country, plus nine lines of craft. And cider. Bucket-loads of cider.
www.fatcatpub.co.uk.

WHAT TO SEE & DO

More stories are told at **Dragon Hall** [9], home of the National Centre for Writing. You can get involved in events and workshops, or just have a look around the medieval merchants trading hall. Clear your mind in **Whitlingham Country Park**, a nature reserve with a couple of lakes in a former gravel quarry in a curve of the River Yare, where you can pick up the **Wherryman's Way** and follow the river to Great Yarmouth.
https://nationalcentreforwriting.org.uk.
www.whitlinghamcountrypark.com.
www.norfolk.gov.uk.

If you see stories as images, one of Britain's most important collections of visual arts is on show at Norman Foster's **Sainsbury Centre** [10], a rare mix of works by Picasso, Bacon and more, and artefacts from Africa, the Americas and Pacific, as well as temporary exhibitions. There's also a Sculpture Park.
www.sainsburycentre.ac.uk.

FURTHER INFORMATION

For more ideas on what to see and do, go to www.visitnorwich.co.uk. Check opening times of beer venues and attractions before you go.

SOUTHWOLD & WALBERSWICK

East Anglia is rich in the soils that the best malting barley grows from, so it's natural to find an appreciation of good ale, not least in the stretch of Suffolk coast, which is home to great traditional pubs and one of the country's most respected family brewers. Southwold, genteel on first appearance, reveals a playful spirit, while Walberswick's dune-backed sands will blow the cobwebs away.

SOUTHWOLD

 ### WHERE TO DRINK

Southwold and **Adnams Brewery** are intimately bound together; it would be unthinkable to visit the Suffolk town without nipping into one of its pubs for a pint, or something to eat. It's a symbiotic relationship that Adnams is keen to make the most of. Stay at one of its hotels locally (see page 143) and you'll be encouraged to extend your visit with a package that includes a brewery and distillery tour, tastings and trips to other attractions. Some will be happy enough to drink the beers close to their source, of course. Ghost Ship pale ale is widely sold across the country, but here you have the chance to try the full range including cask, craft and collaborations with other brewers.
https://adnams.co.uk.

In town, a couple of Adnams pubs stand out. The **Lord Nelson** has been on the beer lover's list of the best place to drink the local brew for many years, and it hasn't changed much with its flagstone floors and open fire, though the beer range has grown somewhat. Next to the Lighthouse, the **Sole Bay Inn** [11] is a more modern operation and acts as Adnams' official tap house, showcasing the entire portfolio including bottles and cans and the latest specials. For beers to take away, along with merchandise and, of course, spirits made at Adnams' own micro distillery, the Adnams store is a pleasant spot to browse and it has a café, too.
www.thelordnelsonsouthwold.co.uk.
www.solebayinn.co.uk.

 ### WHAT TO SEE & DO

Southwold and Adnams are now so intertwined, the brewer even hosts tours of the landlocked Victorian **lighthouse** these days. Beware, if you take the spiral staircase up the 31m (102ft)-tall brick tower there are no intermittent landings on which to regain your puff. But you'll be rewarded at the top by a panorama of the historic streets, the patchwork of **greens** where parts of the town burned down in 1659, the famous multicoloured beach huts that line the groyne-studded sands,

SOLE BAY INN
ADNAMS
SOLE BAY INN
ADNAMS

and the watery battleground where the English
and French fleets clashed with the Dutch at the
Battle of Sole Bay in 1672.
https://adnams.co.uk/collections/tours-experiences.

You'll also spy Southwold's other landmark, the
pier [13], which was rebuilt at the turn of this
century despite the declining popularity of these
structures. But then this one's a bit different, just
like Southwold, where outward appearances can
be deceptive. The retro amusement arcade has
prices to match, and the main attraction, **The
Under The Pier Show**, is an affectionate send-up
of slot machines, featuring Heath Robinson-style
contraptions such as the Micro Break, which, for
just £1, offers an eco-friendly holiday with no risk of
deep-vein thrombosis from the comfort of an
armchair positioned in front of a TV screen.
Normal seaside service is resumed across the road
at the boating pond, which hosts **model yacht
regattas** and the crazy golf course.
www.southwoldpier.co.uk.
www.southwoldboatinglakeandtearoom.co.uk.

Back in the well-heeled town centre, there are a
few gentle diversions, including **CraftCo**,
championing local artists' work, **Buckenham
Galleries**, which shows high-quality
contemporary art, and the curious little **Amber
Museum** – this coast is rich in the resin – in the
back room of a jewellers. You can find out more
about life on the ocean wave, too, at the **Sailors'
Reading Room** [12], which is crammed with
maritime memorabilia and was built to provide
seamen with an edifying distraction from the pub.
www.craftco.co.uk.
https://buckenhamgalleries.co.uk.
www.ambershop.co.uk.
https://southwoldsailorsreadingroom.co.uk.

[13]

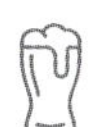

[14]

WALBERSWICK

WHERE TO DRINK

Across the River Blyth, in the village of Walberswick, the **Anchor** [14] is a characterful and colourful retreat run by Mark Dorber – a man regarded as the best cellar manager in the country – and his wife Sophie. As well as a well-kept range of Adnams beers on draught, the fridges here burst with continental classics and American craft, while locally sourced dishes feature on the menu. On the village green, the **Bell Inn** has 600 years of history behind it, evident in the wonky stone floors and curious nooks and crannies. The Adnams range is served at the bar on cask and keg. www.anchoratwalberswick.com. https://bellinnwalberswick.co.uk.

WHILE YOU'RE HERE

The **Star Inn** at Wenhaston is an old-fashioned sort of multiroomed pub with a good line in retro advertising signs and views over the Blyth Valley. House ale Green Jack Golden Best is usually joined on the pumps by at least four others, including from Colchester Brewery, and beer festivals are staged in the garden twice a year. The **Theberton Lion** takes its beers seriously with, typically, a choice of three Anglian-brewed ales plus half a dozen craft keg, again from mainly local brewers. For an unusual final call, **Krafty Braumeister** on the edge of Leiston specialises in brewing in the German style, producing a range of bottled beers. It opens its doors to visitors three days a week for a tour and tasting.
http://wenhastonstar.co.uk.
www.thebertonlion.co.uk.
www.kraftybraumeister.co.uk.

WHAT TO SEE & DO

The point about Walberswick is not the village itself, as endearing as it is with its two pubs, scattering of shops and tea rooms, an art gallery that pops open in the summer, the Heritage Hut with artist John Doman Turner's *Walberswick Scroll*, and some very attractive Arts and Crafts houses (including work by the architect Frank Jennings), where its lucky residents live (especially around the green). The point is what happens where the road ends and the water begins.

If you approach from the road that cuts east from the A12, you have two choices when you reach the centre of Walberswick [15]. Go forward across the slim road bridge, along a path, through a gap in a row of black beach huts, and down a sandy slope between marram-tufted dunes to arrive on the beach braced against the slate grey North Sea [16]. Alternatively, turn left for the harbour, thread your way around the quay where the **rivers Dunwich and Blyth** provide airstrips for passing birds, and go over the hump of dunes to the same stretch of sand and shingle.

The harbour is here because this was an important trading and fishing port from the Middle Ages until the First World War. In fact, in Walberswick's medieval heyday they built a veritable cathedral, the ruins of which surround today's church of **St Andrew's**, suggesting a later bit of belt tightening. But tourism has eclipsed trade, first encouraged by Victorian artists including Charles Rennie Mackintosh and, more recently, the **British Open Crabbing Championships**.

The quirky contest's popularity snowballed until it became a victim of its own success and, officially, doesn't take place now. Its reputation spread so far and wide (helped by T-shirts bearing the slogan 'I caught crabs in Walberswick') that the event

[15]

[16]

overwhelmed this little community. No matter, you can still fish for crabs on **Wally's Bridge**, named after the late Wally Webb, a legendary villager who co-founded the competition (amusingly, he is remembered on blue plaques on houses where he did building work). And you can to-and-fro between Southwold and Walberswick on the ferry – a rowing boat – across the Blyth to Southwold in summer or, year-round, the Bailey footbridge, half a mile upriver.
http://walberswick.onesuffolk.net/heritage-hut/.
 www.solebayteamministry.co.uk/st-andrews-
    walberswick.
www.walberswickferry.com.

WHILE YOU'RE HERE

Head south through the mosaic of woods, heathland and marshes to **Dunwich**, where a city lies beneath the waves, and **RSPB Minsmere**, the domain of red deer, otters, avocets and more.
www.nationaltrust.org.uk/visit/suffolk/dunwich-
    heath-and-beach.
 www.rspb.org.uk/days-out/reserves/minsmere.

FURTHER INFORMATION

For more ideas on what to see and do, go to www.explorewalberswick.co.uk and www. visitsuffolk.com. Check opening times of beer venues and attractions before you go.

HACKNEY

As a hub of urban cool, Hackney has attracted more than its share of craft beer producers and purveyors. Behind the headlines, there's a former rural retreat for nobility, a proud Caribbean heritage and a lot of green space.

HACKNEY CENTRAL

 ### WHERE TO DRINK

Britain's Best Independent Craft Brewery Taproom 2023, awarded by the Society of Independent Brewers and Associates (SIBA), is within a two-minute walk of Hackney Central station. **Hackney Church Brew Co.**, as the name suggests, is a place of secular beer worship. Under the railway arches, the taproom pours an accessible and varied range of fresh beers from the brewery next door. Further down the track, **Deviant & Dandy** brews eclectic, adventurous beers, including one-off specials, served from 10 lines in its brewery and taproom, which doubles as a venue hosting music events and art exhibitions.
www.hackneychurchbrew.co.
www.facebook.com/deviantanddandybrewery

Hackney Tap occupies the Old Town Hall, its stripped-back bar spilling out onto a large, sunny terrace. Up to 20 keg taps are pouring at any one time plus a few from the cask. Nearby, the **Cock Tavern**, original home of Howling Hops, serves eight rotating cask beers plus 16 on keg in a rustic ale house environment. Look out for regular tap takeovers and award-winning pickled eggs.
www.hackneytap.com.
The Cock, 315 Mare Street, E8 1EJ.

Tucked away in the backstreets, the **Chesham Arms** was saved by the local community and, after being closed for three years, reopened in 2015. Comfortable and welcoming with a great garden out the back, its four cask ales come from independent brewers, and there's usually an interesting sour or two on the craft taps. Five Points Brewery took over the spacious **Pembury Tavern** in 2018 and alongside its own range of cask and keg beers you'll find guest options on the 22 lines. There's always plenty going on here, from beer festivals to tastings and educational sessions.
www.facebook.com/thecheshamarms.
www.pemburytavern.co.uk.

WHAT TO SEE & DO

Everyone thinks they know what Hackney is about, but at ground level there are some surprises in

[17]

store. For one it's hard to imagine that this was once a rural retreat for nobility, but the evidence is present in the shape of **Sutton House**. One of the last remaining Tudor houses in London, it was built for Sir Ralph Sadleir, a sidekick of Thomas Cromwell, whose progress up the greasy pole can apparently be deduced by the aspirant choice of brick (some of which is exposed) for a relatively modest home. Step through the front door and Hackney's frantic energy dissipates in the shadowy oak-clad rooms, with exquisite panelling carved like fabric in the Linenfold Parlour, and the open central courtyard.
www.nationaltrust.org.uk/visit/london/
    sutton-house-and-breakers-yard.

Walk through Sutton Place and the old graveyard, its walls stacked with headstones, towards medieval St Augustine's Tower and the shopping street The Narroway. Here you'll find a Turner prize-winning sculpture: three giant representations of Caribbean fruit, Custard Apple (Annonaceae), Breadfruit (Moraceae) and Soursop (Annonaceae), unveiled in 2021 by their creator Veronica Ryan. It's the UK's first permanent public artwork honouring the Windrush generation.

Along the road, Warm Shores by Thomas J Price is a second recent notable sculpture to honour that pioneering generation, two larger-than-life bronzes of a man and a woman created from composites of likenesses of locals connected to the Windrush era [17]. The bronzes stand next to Frank Matcham's **Hackney Empire**, embellished in 2004 with a modern extension, where everyone from Charlie Chaplin to the Rolling Stones have appeared. It stages the best Christmas panto in town.
Warm Shores, Town Hall Square, E8 1ES.
https://hackneyempire.co.uk.

[18]

DALSTON

 WHERE TO DRINK

In a former car park near Dalston Kingsland Station, **40ft**'s award-winning brewery and taproom [18] is constructed out of shipping containers with space for 100 people. Ten taps feature limited-edition brews and collaborations alongside the core range. www.40ftbrewery.com.

KRAFT Dalston is a celebration of German brewing, its modern bar, restaurant and roof terrace serving lagers, weissbiers and pale ales brewed in the basement. Brewery tours are available, led by Kraft's in-house sommelier, where you can learn about what makes these beers different and, of course, taste them. Some of the most highly regarded brewers from around the UK, plus well-chosen imports, are represented on the 20 taps at **Red Hand** amid bare-brick and tile industrial chic.
www.kraftdalston.com.
www.red-hand.co.uk.

 WHAT TO SEE & DO

It was her trips to Dalston's **Ridley Road Market** as a child that inspired Veronica Ryan's Windrush sculpture. Despite rampant regeneration of the area, it remains the beating heart of Dalston. Breadfruit, soursop, custard apples and more can be bought here alongside much else on the cheap and cheerful stalls, but it's the vibe that really draws visitors to this lively, authentic street market. https://hackney.gov.uk/ridley-road-market.

Around the corner is the **Dalston Eastern Curve Garden**, created for locals to enjoy, especially those without an outdoor space. Stop on the way in to admire the **Hackney Peace Carnival** mural, from 1985, depicting anti-nuclear protestors on the march. Hackney Empire is not the only show in town. Dalston's **Arcola Theatre** has won plaudits for championing emerging as well as established talent and hosts the annual opera festival Grimeborne. https://dalstongarden.org.
www.arcolatheatre.com.

LONDON FIELDS

 ## WHERE TO DRINK

Split between a pair of railway arches, **Forest Road Brewing's** impressive, well-appointed taproom serves a beer list dominated by US-inspired IPAs. **Five Points Brewing** moved its brewery into an expansive site on Mare Street in 2021. It houses a 32-hectolitre brewhouse, a barrel-aging store and a splendid taproom with a courtyard where you can drink its full range of modern, reliable, classic beers. Brewery tours and tastings are available. www.taproome8.com. www.fivepointsbrewing.co.uk.

The Dove [19] is one of the great London pubs. For more than three decades this characterful, rambling, multiroomed, wood-panelled local has attracted serious beer lovers with its stock of Belgian brews, now joined on the taps by British craft. www.dovepubs.com.

Not far from Hackney Central, the **Kenton** is a delightfully quirky corner pub run by a Norwegian chap. There's lots of surprising bric-a-brac, a brightly painted beer garden and no fewer than 17 brews on draught featuring local ales, well-chosen craft and its own-brand Norwegian-style lager, Moose Juice. www.kentonpub.co.uk.

WHAT TO SEE & DO

Never mind urban grit, this borough has lots of green space, including several good parks, such as **London Fields**, originally common land. It's home to an Olympic-size **lido**, which first opened in 1932 and after falling into disrepair was rescued and restored in the early 2000s. The park is also the starting point for the **Dunwich Dynamo**, an annual 180km (112-mile) overnight bike ride to the Suffolk seashore that takes place in July. https://hackney.gov.uk/london-fields. www.better.org.uk. www.dunwichdynamo.co.uk.

On the other side of the park is **Broadway Market** where, at number 9, you can still see the gold and green sign of F Cooke's eel, pie and mash shop, the oldest such family-run business in London, which shut in 2019 after 120 years. Today the street draws the crowds for its café society, independent shops and weekly farmer's market – you'll need deep pockets. https://broadwaymarket.co.uk.

WHILE YOU'RE HERE
Walthamstow-based **Signature Brew** has a tap house in Haggerston by Regents Canal. www.signaturebrew.co.uk.

[19]

HACKNEY WICK

WHERE TO DRINK

On the banks of the River Lea, **Crate Brewery** [21] brews off-site now, but you'll find its full range of beers on the bar, typically a couple of pales, a lager, a stout and a sour, plus a couple of changing guest taps and a well-stocked can fridge. With tables along the waterfront as well as inside, there's plenty of space to relax. And if you're feeling peckish, Crate is probably just as famous for its pizzas as its beer.
www.cratebrewery.com.

Next door is the **Howling Hops Brewery** and its taproom, which lays claim to being the UK's first dedicated tank bar. Behind the counter are 10 tanks, pouring beers of almost every style, also available to take away in glass flagons. On a weekday you're very much aware you're drinking in somebody's workplace. At night, though, the space transforms into a high-energy venue.

Opposite you'll spot painted on the wall a bearded gent carrying a placard announcing **Old Street Brewery** [20], a cosier sort of venue with an American vibe. As well as US standards on tap, you might find a saison, a nitro stout or a golden ale with a dash of fruit, plus the odd import from across the pond and a Czech-style lager made by Tottenham's Bohem.
www.howlinghops.co.uk.
www.facebook.com/oldstreetbrewery.

Beer Merchants Tap offers practically any style of brew you can imagine from 20 keg taps, a pair of cask pumps and fridges stocked with some 500 bottles and cans from around the world. If that isn't enough, an on-site blendery ages and blends wild-fermented beers. There are also tutored tastings and training sessions among other beer events.
www.beermerchantstap.com.

THE WHITE BUILDING

👓 WHAT TO SEE & DO

The borough's most famous green space is **Hackney Marshes** at its eastern edge. Here you can stride out under wide skies amid football and cricket matches depending what time of year you visit, snatching glimpses of the City skyline to the south-west. Follow the canal towpath and you'll arrive at Hackney Wick. Along the way, you'll see plenty of graffiti and **street art**, a theme that continues in roads including Bream Street, Fish Island and Stour Road. Notable murals to hunt out include an homage to the Wick's industrial past by Busk at Hackney Bridge on East Bay Lane and two guitarists by Thierry Noir at 117 Wallis Road. Other leftfield pursuits at the epicentre of Hackney cool include **kayaking and paddleboarding** on the canal [22]. https://hackney.gov.uk/hackney-marshes. www.moocanoes.com.

WHILE YOU'RE HERE

Try **Tap East**, an oasis of brewpub calm in the shopping frenzy of Stratford Westfield. www.tapeast.co.uk.

FURTHER INFORMATION

For more ideas on what to see and do, go to www.visitlondon.com. Check opening times of beer venues and attractions before you go.

All Saints Church: Clowns pull on their best baggy suits, loud shirts and frightwigs and head to All Saints Church Haggerston each February to remember the father of clowning Joseph Grimaldi, the original 'joey', born in 1778, before putting on a show for the crowds. www.trinitysaintsunited.co.uk/clown-service.

[22]

WALTHAMSTOW

A straggling industrial estate has become one of the country's most unlikely beer destinations, the Blackhorse Beer Mile. Start at the northern end of the 'mile' so you're close to Blackhorse Road station when you finish. It sits beside the Walthamstow Wetlands, a welcome breath of fresh air in East London, and close to the teenage home of the artist and thinker William Morris.

ON THE BLACKHORSE BEER MILE

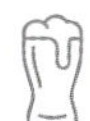

WHERE TO DRINK

At the top of the Mile, founded in 2011, **Hackney Brewery** was one of the pioneers of the London craft beer scene before moving to Blackhorse a decade later. It specialises in accessible hop-forward beers, all available, plus guests, from the 20 lines at the High Hill Tap.
www.hackneybrewery.com.

SIBA Brewery Business of the Year 2023, **Wild Card Brewery**'s Lockwood taproom, wraps around the brewkit with seating on the mezzanine from where you can watch an imaginative variety of beers ferment – and quite often celebrity head brewer Jaega Wise hard at work.
www.wildcardbrewery.co.uk.

A rotating line-up at **Beerblefish Brewing Co.** includes cask beer as well as keg, and there's a family-friendly vibe. It offers brew days when the team will help you create your own beer for that special occasion.
www.beerblefish.co.uk.

Exale Brewery has a gritty, industrial feel while the beers are hoppy, hazy and often experimental in style , showcasing unique flavour profiles. The taproom spills out into the yard and a double-deck roofed seating area, and regularly hosts tastings, live music and quizzes.
www.exalebrew.co.uk.

The newest kid on the block, the understated, **Pretty Decent Beer Co** serves accessible hop-forward beers including lagers, sours and a choice of low-strength alternatives in its bright, airy and welcoming taproom.
www.prettydecentbeer.co.

Easily spotted thanks to the tall, gleaming fermenting vessels that stand in the yard, **Signature Brew** has a purpose-built taproom with a roof terrace alongside the brewhouse. As well as the beer, the big attraction here is the music, pumped out by a state-of-the-art sound system.
www.signaturebrew.co.uk.

[23]

Last stop on the Mile, **Big Penny Social** doesn't have a brewery but makes up for it with the sheer scale of its venue, spread across two massive rooms and an equally huge garden, all welcoming large groups, families and beer lovers alike. Alongside own-brand core beers, interesting guests are served.
www.bigpennysocial.co.uk.

WHILE YOU'RE HERE

Extend your tour by starting at the **Tavern on the Hill**, Wild Card's pub in Higham Hill, where a range of the brewer's beers are available on cask and keg. Wild Card's third Walthamstow venue is the **Barrel Store** on Shernhall Street where it's joined by lager specialist **Pillars Brewery** and its taproom. Pillars also has a bar called the **Untraditional Pub** in the Crate development on St James Street, and was joined there by a second **Pretty Decent Brewery** tap in 2024. Close to Walthamstow Central station, **Collab** is a craft beer and burger restaurant jointly operated by Signature Brew and We Serve Humans.
www.tavernonthehill.co.uk.
www.wildcardbrewery.co.uk.
www.pillarsbrewery.com.
https://prettydecentbeer.co.
www.thecollablondon.com.

WHAT TO SEE & DO

Like Hackney, this London borough has surprises for visitors, including a natural one, the **Walthamstow Wetlands** [23]. The cluster of 10 reservoirs provides water for 3.5 million Londoners and doubles as a nature reserve, where wildlife can thrive and people can walk, cycle, fish and watch the birds chatter on its shores and islands. Herons learning deportment from their parents and great crested grebes tapping beaks in greeting are among other species on view from the well-made paths that meander around the pools of water, past hides and viewpoints. Good use is made of the reserve's industrial scars. Climb to the top of the old Coppermill Tower and look through its Italianate arches across the shining water to Alexandra Palace. There's a viewing deck (cum gallery), too, as well as the café and shop, in the Victorian-era Engine House.
www.walthamstowwetlands.com.

Along the road, another green albeit more cultivated space is **Lloyd Park**, which not only has gardens, a playground and sports courts at visitors' disposal but a moat. The ornamental ditch underscores the rear of an 18th-century

mansion that houses the **William Morris Gallery** [24]. This was the teenage home of the designer, craftsman, writer and radical socialist, a fitting venue for the world's largest collection of the work of the founder of the Arts and Crafts movement. The creative expression of his quest to fuse beauty, functionality and accessibility and his fascination with the medieval age is revealed in the variety of pieces on display, from wallpaper to stained glass, from textiles to furniture. There is work, too, by his close friend the Pre-Raphaelite artist Edward Burne-Jones. www.walthamforest.gov.uk. https://wmgallery.org.uk.

Fast-forward to the future at the gallery of Neon signs, **Gods Own Junkyard**, founded in the 1950s by Dick Bracey and still run by the family. The signs are for sale or rent (they're all the rage with fashion brands and Hollywood), or you can just bathe in their seedy glow in the café. www.godsownjunkyard.co.uk.

Rather more humdrum purchases are on offer at **Walthamstow Market**, Europe's longest (though not the largest) outdoor street market. In the week, it sells commonplace wares. At the weekend, it transforms into the Sunday Social, with artisan makers and bakers selling everything from candles to croissants. More indie and vintage goods can be found at **Wood Street Market** and the **Georgian Village**. www.walthamforest.gov.uk. www.woodstreetindoormarket.co.uk. www.georgianvillage.co.uk.

The ribbon was cut in 2024 on a 24km (15-mile) expansion of Transport For London's Walk London Network. The Green Mile Walk from Peckham to

BEER TALKING

Jaega Wise, Wild Card Brewery

'Walthamstow has to be one of the best places in the country to go get a beer – and I'm not saying that just because I live there!'

Jaega Wise ought to know. She's not only head brewer at Wild Card Brewery, she's also a presenter on BBC Radio 4's *The Food Programme* and Amazon's *Beer Masters* (with singer James Blunt), and author of *Wild Brews*.

She began as a home brewer before training with Sophie de Ronde at Brentwood Brewing and leaving a career in engineering to set up Wild Card with Will Harris and Andrew Birkby. They started in the basement of a pub before expanding into what's now the Barrel Store in 2013 and then the Blackhorse Beer Mile.

'We were the first craft brewer in Walthamstow, and others followed,' she says. 'The attraction is the transport links and the fact the area has an industrial history and the right kind of spaces for a brewery.' https://shop.wildcardbrewery.co.uk/

Epping Forest passes close to the Walthamstow Wetlands along the way. https://tfl.gov.uk/modes/walking/green-link-walk.

FURTHER INFORMATION

For more ideas on what to see and do, go to www.visitlondon.com. Check opening times of beer venues and attractions before you go.

BERMONDSEY

Bermondsey is arguably the most famous craft beer destination in the UK and home to the original Beer Mile, a thread of breweries, taprooms and bars occupying the railway arches under the tracks out of London Bridge into Kent. Here, on the shores of the Thames, you can explore old London, too.

PICK OF THE BEER MILE AND AROUND

 **WHERE TO DRINK**

At last count there were 18 beer venues on the Mile, mingling with many street- food vendors to make this such an attractive day out. To help you home in on a manageable number we've selected six for the beer and another three quirky ones that are worth a look, on a route starting in Enid Street at the Bermondsey Tube end and switching across to Druid Street on the other side of the tracks for the second leg.

One of the first brewers to set up in Bermondsey, **The Kernel** has quietly gone about its business producing consistently great beers earning it the respect of everyone on the scene. Sixteen taps present Kernel's brews at their freshest, mostly hop-forward IPAs and rich dark ales. It's also cracked low gravity table beers, a boon on a long drinking day. www.thekernelbrewery.com.

Manchester brewer **Cloudwater** has its outpost in the Smoke here, a bright and buzzing bar with 20 lines pouring old favourites, new releases and the latest from the brewer's barrel- aging project. It also hosts tutored tastings and bottle-shares of beers beyond its own.
https://cloudwaterbrew.co.

Moor Beer Vaults is next door, with a dozen lines pouring beers from the Bristol brewer, including one on cask. Moor combines modern US brews with traditional British styles, and here you have a chance to try its take on old ale, Old Freddy Walker, on draught. At 7.3% abv, probably just a third, though. There's a live music venue out the back, too. www.moorbeer.co.uk/vaults.

A few arches down, **The Dutch Taproom** is a relative newcomer to the Mile and brings something a bit different. It does pretty much what it says on the tin with 30 taps devoted to beers brewed in the Netherlands, featuring La Trappe and de Molen alongside less well-known brewers such as Uiltje and Jopen, all served in a laid-back atmosphere.
www.pr-dutchdrinks.co.uk/dutch-taproom-it-aint-much-if-it-aint-dutch.

Anspach & Hobday Arch House is on the site of the firm's original brewery. Most production has now moved to Croydon, allowing the space to create a relaxed tap here over two floors. A dozen lines pour a variety of A&H beers plus guests. www.anspachandhobday.com/the-archhouse.

Perhaps the most spectacular venue on the Mile, the **London Beer Factory Barrel Project** [25] specialises in barrel-aged beers and some intriguing creations join regular craft brews on its 24 taps. The walls here are lined with some 200 barrels containing the maturing ales, reaching right up into the mezzanine at the back. It's an atmospheric spot to try something you probably haven't had before. https://thebarrelproject.co.uk.

To add some curiosities to your tour, how about a sip of sake at **Kanpai**, where the fermented rice brew is served in a few different flavours. You'll get a warm welcome and there's a chance to tour the brewery to find out more. Or join the athletes at **London City Runners**, a running club where beer comes a close second. Half a dozen taps pour craft beers and there's a table-tennis table at the back for exercise. And you did remember your wheels, didn't you? Often forgotten at the extreme London Bridge end of the Mile, **Hop Kingdom** is an indoor skate park that's serious about beer, with its own-label pales available alongside some well-known craft names. https://kanpai.london. www.londoncityrunners.com/clubhouse. https://hopking.org.

Thames
BERMONDSEY
ST JAMES'S ROAD
DRUID STREET
TOWER BRIDGE ROAD
ABBEY STREET
ENID STREET
Hop Kingdom Skate Park
Southwark Brewing
Kanpai
London Beer Factory Barrel Project
Billy Franks
Anspach & Hobday Arch House
London City Runners Clubhouse
Hiver Beers
Moor Beer Vaults
Cloudwater
Dutch Taproom
Bianca Road Tap Room
Craft Beer Junction
Enid Street Tavern
Mash Paddle
The Kernel
200 yards

[26]

Off the mile, the **Dean Swift** isn't far away at Butler's Wharf – and it keeps the good beer flowing, from the UK and around the world, from 20 taps.
www.thedeanswift.com.

One of London's most famous pubs, the **George** [26] is the city's last remaining galleried coaching inn. Its maze of bars is dripping with history – beer writer Pete Brown has written an entire book about it. The pub is leased from the National Trust by Greene King, which offers well-chosen guests alongside its own cask ales.
www.greeneking.co.uk/pubs/greater-london/
  george-southwark.

Before the Beer Mile there was the **Rake**. Opening at the turn of the century, this tiny bar (and yard) at the back of Borough Market has championed a great diversity of beer styles and brewers from the start. Its 17 lines, both keg and cask, can include bitter, mild, continental lager, US IPAs, porters, sours and wheat beers. It's a gem.
www.utobeer.co.uk/the-rake.

There are a few decent pubs in and around Borough Market, not least the **Market Porter** among who's claims to fame is that it's interior appeared as the Third Hand Book Emporium in the Harry Potter movie. A dozen handpumps along the bar pull a range of well-kept ales, often including Harvey's Best – unusually common around here thanks to the presence of the **Royal Oak**, Sussex brewer Harvey's only London pub, a beautifully restored Victorian corner house with two panelled rooms either side of a central bar. You can, of course, taste the whole range here, cask and keg – the handpulled Dark Mild is especially recommended.
https://themarketporter.co.uk.
www.harveys.org.uk/pub/royal-oak.

At which point you'll fancy a curry, and the **Gladstone Arms** happens to be a family-run desi pub that has a dozen beer lines pouring rotating cask and craft beers to wash down some terrific Anglo-Indian dishes, including an eastern take on the Sunday roast.
www.thegladpub.co.uk.

Bermondsey can try to divert the eye with its luxury flats, smart offices and galleries, expensive restaurants and pricey cafés but it is the fabric of the place that makes this neighbourhood so special. You feel its history when you walk along streets such as **Shad Thames** [29], cobbles beneath your feet, iron gangways criss-crossing the sky above your head, passing through the shadows of the hulking warehouses where dockers once heaved sacks swollen with tea, coffee, sugar, spices and more. It's Dickens without the squalor. Return to the light by cutting down **Maggie Blake's Cause**, an alley named after a local activist who successfully campaigned to ensure public access to the riverfront by **Butler's Wharf**, the daddy of the warehouses. From here, you could follow the **Thames Path** all the way to Gloucestershire in the west or, more realistically, to Woolwich and beyond in the east. But, for now, just enjoy the close-up of **Tower Bridge** and the shiny architectural assortment box of the **Gherkin**, **Cheesegrater** and **Walkie Talkie**. Further east **Bermondsey Beach** reveals itself at low tide, immediately recognisable from countless TV cop shows as the place where murder victims wash up.
Shad Thames, SE1 2YE.
Bermondsey Beach, Bermondsey Wall East, SE16 4TT.

Back on dry land, the Beer Mile's support system includes a clutch of galleries led by the vast, clinical halls of **White Cube**, part of the contemporary art empire run by Jay Jopling, best known for promoting the Young British Artists. Nearby, **Bermondsey Project Space** is somewhat cosier, a sequence of rooms where artists can mount their own exhibitions, while at **Peter Layton London Glassblowing** you'll be mesmerised by the craftspeople at work in its hot glass studio. There are a couple of offbeat museums, too. The **Science Gallery** at Guy's Hospital fuses science, health and art in its exhibitions, and mount a spiral staircase to the attic of a church tower to find the ghoulish premises of an old apothecary and a **Victorian operating theatre**.
www.whitecube.com.
https://project-space.london.
https://londonglassblowing.co.uk.
https://london.sciencegallery.com.
https://oldoperatingtheatre.com.

To the west, you could pick up a chunk of cheese or a street-food snack in **Borough Market** [27]. Those who enjoy beer's country cousin should spend some time investigating craft ciders at **The London Cider House**. Continue west and the sights get hardcore with the former **City Hall**, **The Shard**, **Shakespeare's Globe** [28] and **Tate Modern**.
https://boroughmarket.org.uk.
www.londonciderhouse.com.
City Hall, The Queens Walk, SE1.
www.the-shard.com.
www.shakespearesglobe.com.
www.tate.org.uk/visit/tate-modern.

FURTHER INFORMATION

For more ideas on what to see and do, go to www.visitlondon.com. Check opening times of beer venues and attractions before you go.

BUTLERS WHARF

ANCHOR AT WALBERSWICK

A stone's throw from bracing coastal walks, the Anchor is in an idyllic spot to get away from it all. The Arts and Crafts-designed main building has four en-suite bedrooms, each with its own view, and there are another six superior cedar-clad chalet-style rooms, three of them dog-friendly, in the garden, for extra privacy. A sumptuous breakfast is included, with the option of swapping locally caught haddock or home-smoked kippers for the full English. The pub menu is based around fresh produce, and the Adnams ales are supplemented by an extensive list of bottled beers and exceptionally well-chosen wines. Visit in August for its beer and oyster festival. www.anchoratwalberswick.com.

THE OLD SHIP

From Hackney Central's high street, you wouldn't really know this Victorian inn is here – you'll have to dip along an alley to find its door. It's rather more obvious from the lane at the back, aka its beer garden, revealing a light and spacious bar arranged with large tables where you can expect a selection including local brews from Five Points, and a tasty menu of wraps, burgers and rotisserie chicken. Upstairs are 10 en-suite bedrooms, six singles and four doubles, done out with a modern simplicity that nods to boutique hotels and sprinkled with a few fun decorations, such as the feature wallpaper showing a donkey being given a drink of beer. These are comfy quarters in the heart of the action, a few steps from the Hackney Empire and just around the corner from the Overground if you want to explore further afield. www.urbanpubsandbars.com/venues/ the-old-ship.

MORE PUBS WITH ROOMS

NORWICH
The Georgian Townhouse [30]
www.thegeorgiantownhousenorwich.com.

SOUTHWOLD
The Swan [31]
https://adnams.co.uk.

The Crown
https://adnams.co.uk.

The Brewer's House
https://adnams.co.uk.

WALBERSWICK
The Bell Inn [32]
https://adnams.co.uk.

The White Hart, Blythburgh
https://adnams.co.uk.

HACKNEY
The Rose & Crown [33]
www.roseandcrownn16.co.uk.

SOUTH

Opposite: Lewes.

BRIGHTON

Brighton is proud of a distinctive beer scene that, for many years, resisted the encroachment of national pub groups. It has its own breweries, its own unique, relaxed vibe, and some terrific pubs. George IV created this, one of Britain's first seaside resorts, with the fantasy palace he built here while a prince. It's a playful spirit that lives on in the rainbow-coloured streets.

AROUND THE STATION & NORTH LAINE

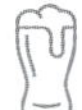

WHERE TO DRINK

Handy for the station, the **Evening Star** [1] is probably Brighton's most famous pub. The original home of Dark Star Brewery, it continues to serve a phenomenal range of beers in a small space. Burning Sky's Plateau is a regular on the handpumps alongside six more ales on rotation, and there are 12 lines of UK craft and continental specialities. Good cider, too. A must-visit. www.eveningstarpub.co.uk.

In Trafalgar Street, which runs under the station forecourt, you'll find the **Prince Albert** [2] thanks to its colourful mural celebrating great musicians. Inside, below the live venue, it's divided into four drinking areas plus a terrace back and front, where you can drink Burning Sky's beers on cask and keg as well as other Sussex brews amid a punky, gothic vibe. The **Lord Nelson Inn** is Harvey's flagship in the city, the original pub now extending all the way to the street corner and to a gallery area at the back. Expect the full range of ales and craft keg from the Lewes brewer. If you can squeeze into the **Great Eastern** – the narrow bar broadens out at the back – you'll have five beers on cask to choose from, mostly local, and a vast range of American whiskeys. https://princealbertbrighton.co.uk. www.lordnelsonbrighton.co.uk. www.greatukpubs.co.uk/the-great-eastern-brighton.

Heading into the depths of bohemian North Laine, **The Pond** is a dedicated craft beer bar with 10 lines that might pour beers from anywhere in the world. It's a cosy spot, and the upstairs room is a good bolthole. **Vine Street Tap** is tiny, and if there's no table available you might politely be asked to leave. But it's well worth a try with at least five very well-chosen craft beers on draught and a fridge packed with cans. www.thepondbrighton.com. www.vinestreettap.com.

Another legendary Brighton pub, the **Basketmakers** seems always to be busy – with good reason. Unusually for a Fuller's house there are eight cask ales to choose from and savour within walls lined

EVENING STAR
ESTABLISHED
THE ORIGINAL DARK STAR 1994 & BREWERY
BRIGHTON SUSSEX
OF THE PREMISES BREWING Co
ALES & STOUTS

1860
PRINCE ALBERT
BRIGHTON
FREDERICK PLACE
THE PRINCE ALBERT
KOPPARBERG

with old tins where you might find a message from some marooned drinker. Around the corner, **North Laine Brewhouse** is a lively modern barn of a brewpub that's part of the Laine pub company. The brewery vessels are visible behind a bar which bristles with 14 lines showcasing the onsite brewer's art in various styles from IPAs to stouts to sours. Tours, guided tastings and brew days are all available to book.
www.basket-makers-brighton.co.uk.
www.northlaine.pub.

In the Open Market on London Road you'll find **The Drop**, a small but perfectly formed craft beer bar and bottle shop operated by Brighton Bier and stocked with a fantastic range of cans and bottles plus a few on draught, all tended by knowledgeable staff.
https://thedrop.beer.

At the **UnBarred Brewery** taproom off Preston Circus you can get up close to the brewing vessels indoors or take a seat in the heated garden at the front. Up to 12 keg lines and one or two cask are dedicated to the firm's own changing range of beers, from easy-drinking pale ales and lagers to sours to the monstrous 10% abv Stoutzilla.
www.unbarredbrewery.com.

A two-bar local with bags of history and architectural interest, and smashing old Tamplin's signage, the **Jolly Brewer Beer House** is worth the short march up Ditchling Road. It's a great supporter of local brewers and you might find

beers from Loud Shirt, Hand and Abyss on three handpumps and a couple of keg lines.
www.facebook.com/thejollybrewerbeerhouse.

WHAT TO SEE & DO

This early British seaside resort is as tempting a prospect today as when George IV, then Prince Regent, romped around it with his mates in the 18th century. Even on a dank day, this place lifts the spirits, largely thanks to paints of every colour that the locals liberally daub on any available flat surface. If the station is your gateway, resist being pulled towards the ocean; instead, turn right out of the main exit (before you reach the gate), then sharp left down Trafalgar Street and, wham, right in front of you, one whole side of the Prince Albert pub is a psychedelic gallery by street artists Sinna One and Req known as *Icons*, depicting the likes of Bob Marley, Jim Morrison, Amy Whitehouse – with an appropriate counterculture signature of kissing coppers by Banksy.
Icons mural, 48 Trafalgar Street, BN1 4ED.

Continue into Sydney Street, Gloucester Road and Kensington Gardens, past tattoo parlours, crystal emporiums, vintage bazaars, comic stores, vinyl outlets, jewellery shops – if there's a non-essential item you wish to buy, you'll find it here. Shift west to the quirky yet quieter streets of Seven Dials, home of the **Dog and Bone Gallery** [3], set in a couple of old red telephone boxes on Powis Square.
www.instagram.com/dogandbonegallery.

[3]

CENTRE

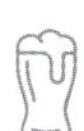 ## WHERE TO DRINK

Hidden behind the seafront Metropole Hotel, in what used to be the city's smallest pub, the **Hole in the Wall** quickly established itself as the go-to spot for Brighton's beer lovers when it opened in 2021. There are two comfortable drinking areas either side of the central bar, and a chalkboard menu displaying the choices of the day, 10 keg lines, four cask and there's always a tap takeover coming up. Nearby is one of the few places to drink good beer on the seafront. **Brighton Bier's** bijoux **West Tap** opens onto the beach and offers six keg lines
and more in the fridge.
www.facebook.com/theholeinthewallbtn.
www.thewesttap.beer.

Off Western Road shopping street, the **Rook Taproom** [4] opened in 2023, sibling to the Independent in Hanover. It follows a similar formula, with 20 keg lines and two cask pouring a range of styles from respected craft brewers such as Beak, Verdant, Rivington and Howling Hops. The overspill room at the back has a Toad in the Hole pub game table, almost obligatory in Brighton's coolest pubs.
https://therook.pub.

Among the eateries of Preston Street, **The Brick** is run by the people behind the Evening Star, but it's quite different, a shiny modern bar specialising in various types of lagers from Czechia, Germany and British craft brewers, sprinkled with a few IPAs among the 20 lines. And there's usually a couple of ales on cask.
www.instagram.com/brick.brighton.

The name of the **Maris & Otter** is a nod to the brewers' favourite variety of barley. This relatively new, smartly furnished Harvey's pub might be the only one with a tap wall dispensing the Lewes firm's

[5]

craft keg beers and one or two guests while the usual Harvey's range is on the familiar handpumps. www.themarisandotter.co.uk.

The **Lion & Lobster** [5] is a weird and wonderful place to explore, pint in hand, over three floors that include a restaurant and a split-level roof terrace. Beware, it's quite easy to get lost in the maze of rooms. Not a huge range of beers but there's always something decent and local on the pumps. https://thelionandlobster.co.uk.

At either end of the city centre you've got a desi destination. **Easy Tiger** [6] occupies the former Hampton pub and alongside the Indian street food there's a healthy range of craft beers on the taps and a couple on cask. The **Permit Room** in the Lanes is a classy yet relaxed joint run by the people at Dishoom. It's open from breakfast time (bacon naan roll, anyone?) and gears up through the day, serving curries and cocktails in the evening along with craft beers from the fridge featuring local brewers such as Beak, Abyss,

Unbarred and Hand. Meanwhile, Cardiff's Mad Dog Brewery opened a taproom over three floors off West Street in July 2024, with a dozen taps pouring its own beers in a wide range of styles. www.easytigerbrighton.com. www.permitroom.co.uk. maddogbrew.co.uk

👓 WHAT TO SEE & DO

There are more colourful shopping streets to the south in the Lanes, particularly known for its cluster of jewellers. The tangle of alleys, or 'twittens', close to the seashore are a remnant of the original medieval fishing village of Brighthelmstone (the **Brighton Fishing Museum** will tell you more). Some say these lanes were paths that fishermen used to reach the patches of ground where hemp was grown, used to make ropes and nets. (Don't be fooled by Dukes Lane, though – it's a reproduction built in 1979.) www.fishingquarter.co.uk.

The Prince Regent set the tone for all the falderol with his **Royal Pavilion**. Even now, even here, the domes and minarets of his Indian-style pleasure palace, designed by John Nash, look a bit trippy. Inside, a social whirl of rooms was designed for drinking, dining, music and dancing. The look is sultan's tent meets Chinese style: richly draped in red and gold silk and velvet, gilded wherever possible, hung with chandeliers of exotic flowers cut from crystal, and embellished with elaborate patterns and scenes from the East, with serpents slithering up columns and through carpets, and dragons peeping from cornices and under mantlepieces.
https://brightonmuseums.org.uk/visit/royal-pavilion-garden.

The bizarre keeps coming down on the seafront with the **Upside Down House** [7], the giant doughnut-shaped bronze, Afloat, and the UFO-style **i360** observation pod that slides up and down a 162m (532ft)-high pole instead of a regulation giant wheel. But it's seaside business as usual on the **esplanade**, which snakes past the bandstand, bars, cafés, ice cream kiosks and fairground rides, and winds up at the triumphantly gawdy **Brighton Palace Pier**.
https://upsidedownhouse.co.uk.
https://www.brighton-hove.gov.uk/libraries-leisure-and-arts/arts-and-culture/public-art-afloat.
www.brightoni360.co.uk.
www.brightonpier.co.uk.

LAGER & WHEAT

HACKER PSCHORR – KELLERBIER	1pt £6.10	5.5% LAGER
HACKER-PSCHORR-MUNICH GOLD	1pt £5.90	5.5% LAGER
PAULANER – MUNCHEN	1pt £5.70	4.9% LAGER
PAULANER – WEISSBIER	1pt £6.10	5.5% WHEAT
LOST & GROUNDED – HELLES	1pt £5.20	4.9% HELLES
LUCKY SAINT	2PT £5.10	0.5% LAGER

CASK

BRIGHTON BIER – SOUTH COAST IPA	1PT £4.60	5% IPA
BRIGHTON BIER – SEAMUS	1PT £4.90	4.7% STOUT
BRIGHTON BIER – ALBERT	1PT £4.40	4.3% NORTHERN BITTER
BRIGHTON BIER – GRACELAND	1pt £5.00	IPA 5.7%
BRIGHTON BIER – PALE	1pt £4.50	4% PALE

KEG

BRIGHTON BIER – FRESHMAN	1PT £6.10	4.5% SESSION IPA
BRIGHTON BIER – PILSNER	1pt £5.50	4.5% PILSNER
BRIGHTON BIER – BIER	1PT £5.50	4% PALE ALE
HANDBREW CO. – OTTO	1PT £5.50	4% DRY STOUT
VAULT CITY – STRAWBERRY SUNDAE	2PT £6.60	5% STRAWBERRY SOUR
BRIGHTON BIER – GRACEKEEPER	⅔ PT £4.60	6% CHOC MILK STOUT
BRIGHTON BIER + LOUD SHIRT – LAST MINUTE AIRPORT GIFT	2/3 £5.00	6.5% WHITE CHOC STOUT
KERNEL – TABLE BEER	1pt £5.50	3.1% PALE
THORNBRIDGE – DDH HALCYON	⅔ PT £5.90	7.4% DDH IPA
UMBRELLA – CIDER	1pt £6.20	5% SPARKLING
ASCENSION – CIDER	1pt £5.50	4% CIDER

HANOVER & KEMPTOWN

WHERE TO DRINK

It's a bit of a trek up to the aptly named **Haus on the Hill** [8] but once there you'll find a friendly local pub with an exceptional range of beers. Typically, 10 keg lines will be pouring big national names such as Deya and Verdant, plus four cask, often Downland from Sussex. And there are, of course, choices from Brighton Bier, which runs this place. Around the corner, the **Independent** [9] curates one of the city's best beer selections with 18 lines of craft, at least two cask and fridges packed with some 200 cans, covering every conceivable style, including a good list of alcohol-frees. Tap takeovers are regularly hosted and there's a Toad in the Hole table at the back. Brighton Bier also has the **Bierhaus** at the bottom of the hill where at least two dozen lines, including four cask, showcase beers from Germany and Belgium as well as UK-brewed craft. And you can browse the fridges for 100 more continental specialities including Trappists, weissbiers and more.

www.hausonthehill.pub.
https://theindependent.pub.
www.brightonbierhaus.pub.

The **Hand in Hand** is a tiny brewpub with a big reputation. Now run by Worthing's Hand Brewery, since 1979, it has housed a unique Victorian-style tower brewery that runs up through the centre of the building over four floors. The resulting beers are served on handpump alongside the Hand core range and guests, plus craft and continental on the taps. Don't look up unless you're not shocked by Edwardian softcore pornography. www.handbrewco.com/venues.

Laid-back and friendly, **The Well** is a beer café and bottle shop in the heart of Kemptown. Just the four craft lines here, but plenty more cans on the shelves. It if looks too busy, there's a 'secret' room behind the bar. www.instagram.com/the_well_brighton.

Part of the 2023 Sea Lanes development, **Bison Beach Bar** [10] is split over two levels with a roof terrace that takes full advantage of a sunny, seafront location. And the beer's good, too, bringing together some of the best-known names in craft with Bison's own-label brews. www.bisonbeer.co.uk/beachbar.

Further out towards the Marina, the **Loud Shirt Brewery** taproom [11] is open Fridays and Saturdays offering eight lines of beer including collabs and guests. It's a bit of a hike but in the summer you can take the **Volks Electric Railway** most of the way from the pier. www.brightonbier.com. https://loudshirtbeer.co.uk. https://volksrailway.org.uk/.

WHAT TO SEE & DO

Kemptown is the epicentre of LGBTQ+ life, a fashionable neighbourhood of Regency squares and crescents, where the shops around **St George's Road** and **St James Street** earn the word boutique. It's a little more sophisticated on the seashore here, too, with its heated, outdoor, Olympic-size swimming pool at **Sea Lanes**, the open-water swimming centre, and a wood-fired **sauna and spa** on the shingle. www.sealanesbrighton.co.uk. https://beachboxspa.co.uk.

WHILE YOU'RE HERE

Just a few selected beer spots in the Hove half of the city. A minute from the station, the **Watchmaker's Arms** is an award-winning micropub. Five cask ales in a variety of styles are served on gravity from the cold room behind the bar, usually including one produced by the pub's own nanobrewery. A couple of craft keg lines, too. **The Urchin** [12] is a real rarity – a brewery and shellfish pub. Half a dozen lines pour its own Larrikin beers in a range of styles that nicely complement seafood. www.thewatchmakersarms.co.uk. www.urchinpub.co.uk.

Closer to the Brighton border, **Bison Beer Crafthouse** is a small bottle shop and bar off Palmeira Square, offering a few well-chosen modern brews on draught, while the **Brunswick**,

[11]

[12]

perhaps best known as a live music venue, is serious about its beer, on both cask and keg, and its large garden is a major draw. There's an escape room upstairs, too. But why would you want to?
www.bisonbeer.co.uk/hove.
www.brunswickpub.co.uk.

A new 53km (33-mile) section of the England Coast Path between Shoreham-on-Sea and Eastbourne was opened by Natural England in 2022. Use it to extend your wanderings to Hove and beyond.
https://englandcoastpath.co.uk.

FURTHER INFORMATION
For more ideas about what to do and see, go to www.visitbrighton.com. Check opening times of beer venues and attractions before you go.

Brighton & Hove Pride: Britain's gay city celebrates the LGBTQ+ community with a huge four-day city-wide annual party, now quite a different kind of affair from the original demonstration for gay rights in 1972. www.brighton-pride.org.

LEWES

For many years beer in Lewes meant one thing – Harvey's. But a couple of new craft breweries have brought a greater diversity to the town. It's a terrific, compact, destination for a pub and taproom tour. While you're here, find out why this little Sussex town has associations with some big names, including Anne of Cleeves and Thomas Paine.

 WHERE TO DRINK

Perhaps the best place to start is the bridge, with its fine view of **Harvey's Brewery**, the oldest in Sussex. Founded in 1790, the family brewer still stands resolute on the banks of the River Ouse and its beers are widely available here. Brewery tours are offered but there's a long waiting list, so book early. Meanwhile, browse the well-stocked shop for Harvey's beers and merchandise. And across the road, down a side street, there's the **John Harvey Tavern** [13], the brewery tap, where you'll find the full Harvey's draught range, some served straight from the cask, unusually.
www.harveys.org.uk.
www.johnharveytavern.co.uk.

In the town's old brewing quarter behind Harvey's, **Abyss Brewing** offers a complete contrast. The taproom is in the middle of the brewery, and you can enjoy a variety of cutting-edge modern beers with the gleaming tanks looming over you. There are tasting events to hone your palate, and Abyss – named after the cellar of the **Pelham Arms** where it was born – is home to its own cycling club.
https://abyssbrewing.co.uk.
www.thepelhamarms.co.uk.

Further out, via a walk along the river, Only With Love Brewery has set up shop at the **Brewcafe** in Malling Community Centre. From this green and sunny spot overlooking playing fields it serves its craft beers alongside tea, coffee and cake.

Beak Brewery, on an industrial estate across the A26 beneath Lewes's dramatic chalk cliffs, is one of the most celebrated additions to the British craft beer scene in recent years, specialising in delicious pale ales at all strengths. Its taproom, with 15 keg lines, spills out onto a yard that hosts everything from DJ sets to weddings, and look out for Beak's These Hills festival in the summer.
https://beakbrewery.com.

Pop into the **Snowdrop**, a pub named after a freak avalanche that, in 1836, killed 15 people. This friendly, quirky freehouse has a decent choice of beers on cask and keg. Back on Cliffe High Street, the **Gardener's Arms** is a long-time favourite with cask ale aficionados. There's always a good variety on the pumps, served with care. A bit basic, but it's the beer that counts here.
www.facebook.com/thesnowdropinn.
www.facebook.com/GardenersLewes.

LEWES BEER MILE

In the spring of 2024, **Only With Love Brewery** launched the Lewes Beer Mile in collaboration with the town's three other brewers. Those who complete the route can collect a stamp at each brewery and claim a free tote bag. Although you can do it in any direction, the official start is Only With Love's **Brewcafe** in the Malling Centre, a joint venture with Lewes Town Council that acts as a community hub – as well as serving good beer alongside the tea and cakes.
www.onlywithlove.co/pages/lewes-beer-mile.
www.onlywithlove.co/pages/lewes-brewcafe.

Offering something rather different, **The Patch**, on Market Street, is a relative newcomer, a beer café that's open from breakfast time with 10 keg taps featuring some of the big names in UK craft brewing – and often one or two rarities. **The Rights of Man** [15] is also a recent addition, transformed by Harvey's into a comfortable multiroomed high street pub with a roof terrace. A good spot to sample the brewer's range.
www.thepatchbeercafe.com.
www.rightsofmanlewes.com.

Tucked away behind the castle on Mount Place you'll find the **Lewes Arms** [14], famous for a drinkers' rebellion that forced the pub to keep serving some of the best Harvey's Best in town. An unspoilt wood-panelled gem now owned by Fuller's, it's full of character and home to the world pea-throwing championships. If you're on the western side of town, a few doors past Anne of Cleves House there's the **Swan Inn**. Always a good pint of Harvey's pouring here, and the grub's got a strong following, too.
www.lewesarms.co.uk.
https://theswaninnlewes.co.uk.

RIGHTS
OF
MAN
THOMAS PAINE
1791
HARVEYS of LEWES
RIGHTS OF MAN
HARVEYS
FINE SUSSEX ALES
HARVEY & SON LEWES LTD.
FOUNDED 1790

WHAT TO SEE & DO

For a small town, some big names are associated with Lewes. The hobnobbing began with the arrival of William de Warenne, 1st Earl of Surrey, who had fought with his mate William at the Battle of Hastings. He built **Lewes Castle** [16] to guard his territory. Above the bailey rise two stone towers, the original entrance and a 14th-century barbican gate. Take the steps up to a second motte and the South Tower for changing perspectives of the twin defences and balcony views of the Sussex Downs. Back at ground level, the **Barbican House Museum** gives a thorough account of the fort and the history of the local area from prehistoric times.
https://sussexpast.co.uk/attraction/lewes-castle.

In **Priory Park** the romantic ruins of the 12th-century Priory of St Pancras reveal that the influential French monks of Cluny decided to build their first monastery in Britain here. You can trace the church, watermill, bakery, brewhouse, dormitory and infirmary, but the largest surviving part is a schoolboy rib-tickler, the toilet block. The modern sculpture of a helmet in Priory Park is a memorial to the Battle of Lewes of 1264, when Simon de Montfort briefly seized power from Henry III. Some of the Caen stone from the Priory was used on the 16th-century manor house in nearby **Southover Grange Gardens**.
www.lewespriory.org.uk.
Southover Grange Gardens, BN7 1TL.

Next up, the fourth wife of Henry VIII. **Anne of Cleves House** [17] was part of her annulment settlement although she never stayed here. It's a fine example of a 15th-century Wealden hall house with a pretty garden that reveals Tudor planting tastes, and a kitchen and parlour with an original hearth from the period.
Anne of Cleves House, https://sussexpast.co.uk/attraction/anne-of-cleves-house.

But this town fair bristles with pride at its association with the revolutionary activist and philosopher **Thomas Paine**, the author of *The Rights of Man*, who played active roles in the American and French Revolutions. He lived at Bull House from 1768 to 1774 and wrote his first political pamphlet here, *The Case for the Officers of the Excise*. The house reopened to the public in the

Lewes Football Club: When Lewes Football Club became the first in the world to pay their women's team the same as its men's team the news hit the headlines. That was 2017 and at time of writing they remain alone in this commitment to gender equality. It's fitting, then, that the ground was chosen by the artist Amanda Cotton in 2022 as the spot for her sculpture *Inexorable*, dedicated to the trailblazing 18th-century female pirates Mary Read and Anne Bonny.
https://lewesfc.com.

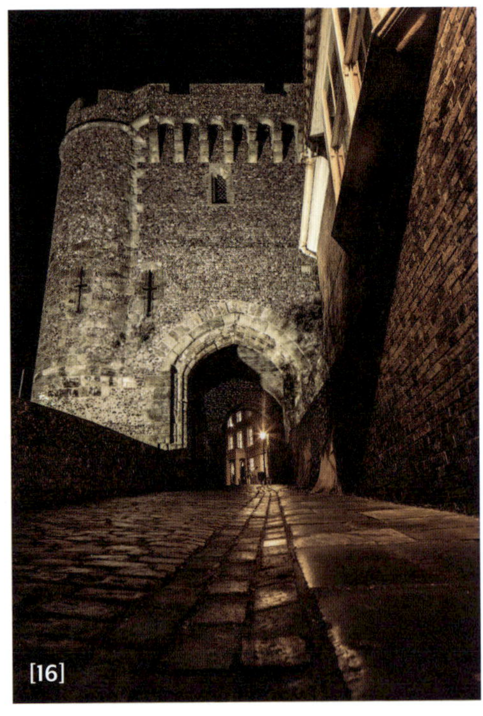
[16]

[17]

summer of 2024. He also has a statue outside the library, a mural in the Market Tower passage, he appears on the local currency, the Lewes Pound, and on Harvey's Tom Paine Ale, and his most famous work is the name of one of the town's pubs.

WHILE YOU'RE HERE

More names to drop, Lewes is within easy reach of the opera house at **Glyndebourne**, and **Charleston**, country retreat of the Bloomsbury Group. www.glyndebourne.com/. www.charleston.org.uk/.

FURTHER INFORMATION

For more ideas on what to see and do, go to www.visitlewes.co.uk. Check opening times of beer venues and attractions before you go.

Lewes Bonfire Night: Lewes holds stages the UK's biggest Bonfire Night celebrations, with thousands of spectators converging on its streets on 5 November. It's also one of the most controversial, marred in past years by the use of blackface and religious antagonism. www.lewesbonfirecelebrations.com/

CANTERBURY & FAVERSHAM

This is hop country and it's also the home of Shepherd Neame. A short train journey joins the brewing town of Faversham with Canterbury to give the travelling beer drinker a variety of options. Along the way discover a fine old market town with an explosive history and a mighty cathedral where pilgrims and visitors mingle.

CANTERBURY

 ## WHERE TO DRINK

On the outskirts of the city, not far from Canterbury East station, **Floc Brewery** has plenty of craft credentials producing an array of hop-forward beers that you'll see at bars in the know across the country. Only here, though, can you find a dozen of them together. The taproom on the brewery floor opens at weekends and there's a fun feel to the whole enterprise. https://flocbrewing.com.

Canterbury's other brewery is at the **Foundry** [18], home of **Canterbury Brewers & Distillers**, an old industrial space converted into a comfortable pub and restaurant. A remarkably wide range of beers are produced on site, served on cask and keg, including rye ales, nitro stouts, wheat beers and lagers alongside the pales and IPAs. Tours and tastings available.
https://canterburybrewers-distillers.co.uk/
   the-foundry-brewpub.

Near the Cathedral, the **Bell & Crown** is a traditional red-brick local specialising in locally brewed cask ales, while around the corner

Shepherd Neame's **Parrot** talks eloquently of ancient times in a lavishly beamed building that dates back to the 14th century. There are six handpulled beers on the bar.
www.facebook.com/bellandcrowncanterbury.
www.parrotcanterbury.co.uk.

Further along on Northgate, the half-timbered **Thomas Tallis Ale House** sings a hymn to good beer in atmospheric surroundings with 12 rotating keg lines showcasing some of the best UK craft brewers plus a couple of cask options. There's no bar counter, just order from whoever looks in charge.
https://thethomastallisalehouse.co.uk.

Mid-terrace among the backstreets, the **New Inn** is worth seeking out, a friendly local with up to seven beers on cask, mostly from Kent and the surrounding counties. Beer festivals are staged twice a year.
www.newinncanterbury.co.uk.

On busy St Dunstans Street, the **Pegasus Tap Room** offers a welcome escape for the beer lover,

The
FOUNDRY
FOUNDRY

rivalling the Thomas Tallis for range with 14 lines of craft. Brewers such as Floc, Verdant, Azvex and Cloudwater are typically represented. www.pegasustaproom.com.

👓 WHAT TO SEE & DO

You're not the first to make a pilgrimage to Canterbury. The Romans tramped here from halfway across Europe in the 1st century – well, hardly a pilgrimage; they came, saw, conquered and called the place Durovernum Cantiacorum. The true faithful pilgrims started coming to the **Cathedral** to honour the archbishop Thomas à Becket a couple of years after he met a grisly end in 1170 – his name is carved on the floor at the spot where he was murdered – encouraged by claims of miracles being performed at his shrine in the Trinity Chapel. The most famous pilgrims are the ones imagined by Geoffrey Chaucer, whose **statue** stands on a plinth decorated with characters from his Canterbury Tales, though their faces are of local people, including the actor Orlando Bloom. Chaucer's bawdy crew took the **Pilgrim's Way** here from Southwark, which you can follow, too, as well as the path from Winchester, the **Augustine Camino** from Rochester or Ramsgate, and the **Via Francigena** from Rome. All roads lead to Canterbury. www.canterbury-cathedral.org. Geoffrey Chaucer statue, 23 High Street, CT1 2AY. https://britishpilgrimage.org. www.viefrancigene.org.

Today's pilgrims are the millions of tourists who visit each year, still largely here to see the impressive seat of Anglicanism, a cathedral established in the 6th century. Once here, they and you can take a shorter walk within the city's walls to see all the other main sights: the medieval streets of the **King's Mile** [19] and its 17th-century **Crooked House**, which looks at perpetual risk of keeling over; the medieval gateway **Westgate Towers**, where you can stand on the castellated battlements and survey the scene below; and the pretty River Stour meandering through the city's streets, on which you can take a **guided tour in a punt** [20]. https://thekingsmile.org.uk. Crooked House, 28 Palace Street, CT1 2DZ. www.onepoundlane.co.uk/westgate-towers. www.canterburypunting.co.uk.

The well-curated **Beaney House of Art & Knowledge** is worth a quick spin – it's to be hoped the families of Oliver Postgate and Peter Firmin will leave the Smallfilms gallery in place here, filled with memories of the children's TV classics they created just down the road at Blean, including *Bagpuss*, *The Clangers*, and *Noggin the Nog*. You might leave Canterbury as you came but you could go onwards along the old train track to Whitstable, the **Crab and Winkle Line**, which locals claim was the world's first regular passenger railway service. It's now a pleasant greenway to walk and cycle, perhaps armed with a picnic from the farmer's market in **The Goods Shed**. https://canterburymuseums.co.uk/the-beaney. https://crabandwinkle.org. https://thegoodsshed.co.uk.

[19]

[20]

FAVERSHAM
WHERE TO DRINK

Faversham has a long brewing history, and **Shepherd Neame** [21] is at the heart of it. The family firm is well prepared to welcome guests at its visitor centre next door to the brewery, where you'll find a shop and a pub. It's also the starting point for one of the country's best brewery tours where you can really get a sense of history as well as a grasp of the brewing process – and there's a tutored tasting at the end. It also hosts an evening tour with an Ale Sampler's Supper and beer-and-food-matching dinners.

Sampling the beers made here, from the well-known core ales and Cask Club specials to the hoppy modern Bear Island brews and bottled classics, can continue in one of the company's nine pubs in town. The pick, both close by, are probably the 14th-century **Sun Inn**, with its oak beams and inglenook fireplaces, and a comfortable corner at the multiroomed **Bear Inn**. www.shepherdneame.co.uk.

www.sunfaversham.co.uk.
www.bearinnfaversham.co.uk.

When you've had enough of Sheps beers, **Furlongs Ale House** is Faversham's first micropub and has a strong local following with eight lines equally split between cask and craft. There's a beer garden out the back. **Creeker's Tap** is a comfortable, two-roomed bar with fish-themed decor in a former glazier's shop. Run by **Whitstable Brewery**, it has 20 draught lines, including two cask, and a beer range extending well beyond the brewer's own. www.facebook.com/FurlongsAleHouseFaversham. www.whitstablebrewery.co.uk/taprooms.

Much-admired by the local cask ale cognoscenti, the **Elephant** is just the other side of Faversham Station. Its five handpumps pour a changing range in a good choice of styles. Elephant, 31 The Mall, ME13 8JN.

Some towns are a pleasure to walk around and Faversham is just such a place. Pick up one of the well-composed **town trail** leaflets from the **Visitor Information Centre** (where you can also view the town's medieval version of the Magna Carta) or **Fleur de Lis Heritage Centre** and set off to see Faversham's physical heritage. Among the points of interest are the **Guildhall**, held aloft on pillars, from Elizabethan times, the medieval Town Warehouse, later named **TS Hazard** after the ship Faversham contributed to the fight against the Spanish Armada, and the colourfully embellished **Shepherd Neame offices** on Court Street, with a hop motif creeping up the columns flanking the front door.
https://favershamsociety.org.
www.favershamcharters.org.

At this point you might feel the urge to abandon counting off the sights so slavishly because ahead lies **Abbey Street**, which deserves no distractions. Grand processions used to advance down this road to the monastery until it was dismantled by Henry VIII. Fortunately, later attempts to erase the past have been resisted and the gallery of medieval and Georgian merchant houses, cottages and remnants of the abbey remain. They're hewn from stone, half-timbered, and built brick by brick, decorated with flint or covered with render, painted in fetching pastel shades or weatherboarded against the elements. They come in all shapes and sizes: tall, short, stout and petite, standing proud or sagging under the weight of time, gable windows peeping out of roofs with tiles curling at the edges like stale toast.

Abbey Street will deliver you to **Standard Quay** [22], a tidal creek once busy with boats ferrying goods to and from London and Europe (this town is terribly well connected – it's also on the Roman road between London and Dover). Now river boats sit silent on their moorings and the Georgian and Victorian warehouses are used for cafés, galleries and antiques shops. Such an easy transport link contributed to the rise of the local brick industry, which built much of Victorian London, and the manufacture of explosives on Faversham's wetlands. At **Chart Gunpowder Mills** further along the creek, credited with helping win

KENT – THE GARDEN OF HOPS

Hops have been grown in this part of Kent for centuries, and the distinctive angled chimneys of the oast houses where they are traditionally dried still punctuate the skyline. The craft beer revolution has brought an explosion of new hop varieties around the world, but Kent has not been entirely left behind. New varieties such as Harlequin, Jester and Ernest have been bred to bring a fresh range of flavours to hop-forward beers, and in 2023 the National Hop Collection, with the support of Shepherd Neame, was transferred to larger premises at Ospringe, near Faversham. It's a museum of some 360 varieties of hops, and there are more on the way.
www.britishhops.org.uk.

the battles of Trafalgar and Waterloo, you can learn more about those boom times.
https://standardquay.co.uk/.
https://favershamsociety.org.

FURTHER INFORMATION

For more ideas on what to see and do, go to www.canterbury.co.uk/, www.visit-swale.co.uk and www.visitkent.co.uk. Check opening times at all beer venues and attractions before you go.

REPERTOR HARWICH
HENRY
HENRY

ISLE OF THANET

The Isle of Thanet, which hasn't been an island for some 350 years, offers beer lovers the highest concentration of micropubs in the country. We're focusing on three seaside resorts, each with its own distinct character, that are within minutes of each other thanks to the train line out of St Pancras International.

MARGATE

 ### WHERE TO DRINK

Margate is the young and lively resort. Most of the best beer is in the Old Town. **Fez** [23] is a must-visit, a micropub done up as a bizarre bazaar. Every inch of the walls is smothered in old advertising signs, brightly coloured bric-a-brac and memorabilia. You could spend hours here just looking – the best vantage point being a repurposed waltzer car in the window. There are usually a few well-kept ales to choose from, poured direct from the casks, and music from the pub's vinyl collection, or occasionally live. Fez, 40 High Street, CT9 1DS.

The Two Halves is a more conventional micropub looking out over the harbour. Local ales are served straight from casks behind the bar – often by the owner, who might look familiar. His day-job is working as a Harry Redknapp lookalike, another reason for calling this the Two Halves, perhaps. Right next door there's a very different sort of beer venue – the **Xylo Microbrewery & Tap Room** [24]. It's a relaxed, welcoming modern bar with a dozen keg taps serving mostly beers brewed on the premises, including lagers, wheat beers and IPAs, plus a few well-chosen guests. There's a nod to tradition, too, in a couple of shove-ha'penny boards. The game seems quite popular in these parts. And you can bring your own food. **Little Swift** is a bohemian sort of place with the feel of a café-bar, serving up to 12 keg beers from acclaimed brewers around the country alongside wines, cocktails and a strong selection of ciders. Two Halves, 2 Marine Drive, CT9 1DJ. https://xylobrew.com. www.facebook.com/LittleSwiftCT9.

[23]

XYLO
MICROBREWERY
& TAP ROOM

On the outskirts of the Old Town, the pretty Grade II-listed **George & Heart** [25] opened its doors as a food-led pub-hotel in 2023. It's serious about beer, too, and has a few craft choices on the keg lines as well as its own-label cask ale, brewed by Kent's Northdown.
www.georgeandheart.com.

Take a stroll along the town's harbour wall and you'll find the **Harbour Arms**, a rustic micropub with a nautical theme and a great panoramic view back across the water. A modest range of cask ales are on offer, which appear mysteriously from a room at the back. But it's a cosy, quirky spot well worth a visit.
www.facebook.com/harbourarms.margate.

Margate is also home to **Ales of the Unexpected**, a celebrated micropub that's a little walk away on the western side of town, past the railway station. It's a no-frills bar that's all about the beer with, typically, four cask ales sourced locally and further afield.
Ales of the Unexpected, 105 Canterbury Road, CT9 5AX.

Halfway back from here to the station you can stop off at the **Shakespeare**, which reopened in late 2023 after being closed for eight years. There's now a decent range of cask and keg beers on the bar, including a surprise few from Five Points Brewery. Turns out the owner, an architect, shares office space with the brewer in Hackney.
Shakespeare, 1 Canterbury Road, CT9 5AQ.

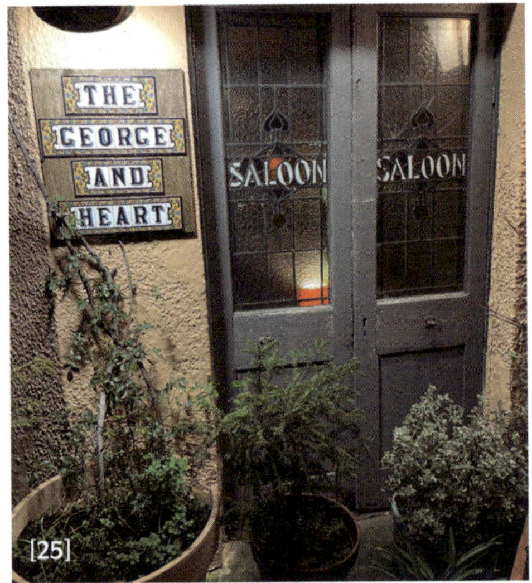

[25]

WHAT TO SEE & DO

Margate has found its place in the 21st century. Thanks first to JMW Turner and, latterly, Tracey Emin, the seaside resort is more paintbrush and easel than bucket and spade these days. The obvious place to go first is the **Turner Contemporary**, six sharp-edged white boxes with sloped roofs that draw the eye to the eastern end of the town's sweep of golden sand. Inside, it's hard to know which way to turn – look at the exhibits, a regularly changing programme of historic and contemporary art, or out to sea through the vast windows and appreciate the special light that kept the gallery's namesake returning to the north Kent coast. One exhibit you can see through the window is Sir Antony Gormley's Another Time, a trademark cast-iron figure standing in the sea. Emin championed her hometown before and after the gallery opened in 2011, opening her own **TKE Studios** in an old bathhouse in 2023, creating workspaces for professional and aspiring artists and galleries showing regular exhibitions.
https://turnercontemporary.org.
www.traceyeminfoundation.com/tke-studios.

There is more art popping up around town, especially in the fast-gentrifying streets around Market Place and King Street, where the influential curator **Carl Freedman** opened a gallery in 2019, and Northdown Road in up-and-coming Cliftonville. This optimism is catching. Even the amusement park **Dreamland**, long a dilapidated reminder of the days when holidaymakers made for the coast rather than the airport, is back open and its latest recipe of vintage rides and star-studded live music seems to be keeping the visitor numbers up. Who knows, perhaps **Arlington House**, the Brutalist tower block by the station, and the abandoned Art Deco **Cliftonville Lido** will be up for some overdue attention soon.
https://carlfreedman.com.
www.dreamland.co.uk.

WHILE YOU'RE HERE
Exercise the little grey cells at the **Crab Museum**, where the crustacean is a vehicle for the bigger questions about life. Hornby, Scalextric, Airfix... **The WonderWorks** is the hobbiest's dreamland, with free workshops and game nights as well as exhibits.
www.crabmuseum.org.
www.wonderworksmargate.co.uk.

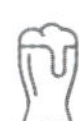

[26]

BROADSTAIRS

WHERE TO DRINK

The smallest and quietest resort of the three, Broadstairs enjoys way more than its fair share of micropubs, in several varieties. Near the seafront, **Sonder**, formerly known as the Thirty-nine Steps, is among the grander sort of the genre, being quite spacious, with a mezzanine for extra tables. The beer list, too, is extensive. Eighteen keg lines feature a range of British craft and continental brews, alongside half a dozen handpumps for cask. Sonder, 11–13 Charlotte Street, CT10 1LR.

The Magnet [26] is named after the comic which, before the First World War, was home to Billy Bunter and his private school chums. The pub is much more inclusive, a place for the whole community, where cask and keg beers are served, with an emphasis on the local. Opposite is **The Chapel**, which doubles as a bookshop over two floors in what is reputedly the oldest building in the town.

It also has a mix of craft and cask.
www.facebook.com/TheMagnetMicropub.
www.facebook.com/TheChapelBroadstairs.

There is another cluster of micropubs around the railway station. Furthest out is the **Four Candles**, which claims to be the smallest brewpub in the country and is frequently crammed with people drinking the ambitious range of ales brewed in its cellar. **The Royston** serves its ales straight from casks housed in a refrigerated cabinet in the middle of this friendly pub and likes to vary the small range with interesting selections from around the UK. Excellent locally made pork pies are also available. And in front of the station there's the railway themed **Mind the Gap**, split over two levels and featuring regular live music.
https://thefourcandles.co.uk.
www.facebook.com/theroystonmicropub.
Mind the Gap, 156 High Street, CT10 1JA.

Don't confine yourself to the town's boundaries; Dickens loved to yomp across the surrounding countryside and regularly made the journey on foot from **Broadstairs** to **Ramsgate**, a route you can follow across the clifftops or at shore level. Alternatively, turn north towards Margate to see some of Kent's finest scoops of sands, including **Joss Bay**, **Stone Bay** and **Botany Bay** [27] with its spectacular chalk stacks. Return to Broadstairs and reward yourself with an ice-cream from **Morelli's**. www.morellisgelato.com.

While Margate is all about visual art, Broadstairs is all about the written word. This is where **Charles Dickens** spent his holidays for more than 20 years in the mid-1800s, during which time he worked on novels including *Nicholas Nickleby*, *The Old Curiosity Shop*, *Barnaby Rudge* and *Martin Chuzzlewit*. He finished *David Copperfield* at Fort House, since renamed **Bleak House**, where you can check in for bed and breakfast, and Copperfield's aunt, Betsey Trotwood, was inspired by a local woman whose former home on the seafront is now the **Dickens House Museum**. Here you can also pick up a town trail leaflet and go about the winding streets lined with pretty flint buildings to discover the Broadstairs he knew. http://bleak-house-guest-house.hotelskent.com. www.visitthanet.co.uk/attractions/dickens-house-museum-1999.

RAMSGATE

 WHERE TO DRINK

The largest town of the three, Ramsgate is the home of Thanet's first micropub, the **Conqueror Alehouse** [28] (the Butcher's Arms lying just the wrong side of the River Wantsum). A long, narrow room made narrower by the fridge from where the cask beers are served (there's no bar counter), it's comfortable and friendly with a good choice of ales. It's now run by the family that owns the **Artillery Arms** [29] around the corner, a splendid old pub that also cares about its beer with a constantly varying range on the pumps, including from craft brewers. There's a military theme, but in a quaint, almost camp, way.

Conqueror Alehouse, 4C Grange Road, CT11 9LR.
Artillery Arms, West Cliff Road, CT11 9JS.

[29]

Eddie Gadd, Ramsgate Brewery

'Thanet has gone from a desert to an absolute oasis for beer. If you're looking Kentish beer, we have it all here now.'

It's been 'a stunning turnround' since Eddie Gadd first arrived in Ramsgate on a wet Tuesday in February 2000. The Firkin chain brewer had been invited to set up his own brewpub in town, 'and I fell in love with the place'. Six years later he expanded into a standalone brewery and started to sell his beers all over the island. A visitor can barely walk up to a bar and not see Gadd's name on the pumps.

'I was told nobody drinks real ale here. There were only three or four independent pubs selling it, but that changed when we launched the first Easter Beer Festival in 2006. It turned out they loved cask beer, they just hadn't had the opportunity to try it.'

The explosion of micropubs that followed the opening of Thanet's first, the Conqueror in Ramsgate, in 2010, then brought a huge diversity of cask ales to the area and created the great beer destination it is today.

As for Eddie, he's now working with Kent's hop breeding programme, trying to find the genome sequence that guarantees a tasty beer!

www.ramsgatebrewery.co.uk/

THE UK'S FIRST MICROPUB

It was in 2005 that Martyn Hillier launched the **Butcher's Arms** in the village of Herne, inspiring a movement that's spread across the country and not only generated a new species of local pub but provided many small brewers with essential outlets for their ales.

While the Butcher's Arms technically isn't in Thanet, it's the reason so many micropubs have opened here in neglected shops and given us a fantastic showcase of the way Hillier's simple formula, a one-person operation serving only cask beer, has evolved into an imaginative array of drinking places.

www.facebook.com/Micropub/?locale=en_GB.

In the centre of town two more micropubs are worth a look. The **Hovelling Boat Inn** (a hoveller was a kind of freelance seafarer) has up to half a dozen cask ales served from a cold room and a pétanque court out the back, while **The Pub** belies its simple name by specialising in Bavarian brews – as well as having three cask beers on handpump. Hovelling Boat, 12 York Road, CT11 9DS. https://tprtest.dreamhosters.com.

In the backstreets to the east, it's worth seeking out the **Montefiore Arms** [30], a full-sized three-bar pub that acts as a tap house for Gadds beers from the Ramsgate Brewery. As well as a great range of cask ales, including guests, there's usually

a couple of rare Gadds on the keg taps. And nearby, you'll find the 17th-century **Honeysuckle Inn**, a characterful retreat with some good local beers from the cask and the keg, plus live entertainment. https://montefiorearms.co.uk. www.facebook.com/honeysuckleinn.

On the outskirts of the resort lies **Ramsgate Brewery**, source of the Gadds beers you'll see all over east Kent, but only rarely outside. It has a shop and a taproom, and you can call direct to arrange a brewery tour. **Wantsum Brewery** is another name you'll frequently see on pump clips. It's based on the western coast of the 'island' and offers tours and a taproom. Thanet is also home to the smaller **Northdown Brewery**, with a tap opening limited hours. www.ramsgatebrewery.co.uk. https://wantsumbrewery.co.uk. www.northdownbrewery.com.

The Napery: The Walpole Bay, an Edwardian hotel in Margate, is the unusual setting for a unique art gallery, which has been exhibiting linen napkins decorated by guests since a visitor sketched a scene on one and presented it to the owner, Jane Bishop, as a 'magical memory' of his stay in 2009. www.walpolebayhotel.co.uk/napery/.

👀 WHAT TO SEE & DO

It's a 10-minute walk downhill from Ramsgate's railway station to the town centre, and your opinion of the place might follow a similar trajectory as you pass through its drab shopping streets. Persist; there is a beautiful waterfront just ahead filled with yachts that have found a convenient berth just 30 nautical miles from the French coast. Ramsgate was a key port for the nation from the Napoleonic Wars to the evacuation of Dunkirk in 1940, and until recently a place to hop on a ferry to the Continent (the resumption of such a service was under discussion at time of writing). These stories will be told when the **Ramsgate Clock House** [31] – formerly the Maritime Museum – reopens as a heritage centre following a major renovation that will also turn the **Pier Yard** into a space for markets and events.

There are a few sights dotted about the harbour – you could look the **Obelisk** up and down (a thank you to George IV for making this Britain's only royal harbour), circumnavigate the squat Georgian **lighthouse** [32] on the pier, and wonder at the sign on one building that says 'The Ramsgate Home for Smack Boys' (a home for the apprentices of fishermen who worked on boats called 'smacks'). No yacht required, you can even sail along the coast on the **River Runner**. https://ramsgateroyalharbour.co.uk. https://river-runner.co.uk.

By the **Sailors' Church**, a steep staircase zigzags up the wall – climb it to fully appreciate its name, **Jacob's Ladder**. Turn left at the top to join the **Westcliff Promenade**. If you thought the sea views were great from the harbour, you should take them in from up here on the cliff. Turn up Screaming Alley to St Augustine's Road to see the **church designed by Augustus Pugin**, the architect responsible for the proliferation of neo-Gothic buildings across the country. He also built himself a home here, **The Grange** (where you can stay), with a tower from which he would look out to sea on rough days and jump in his lugger, *Caroline*, if shipwrecked sailors needed to be rescued. www.augustine-pugin.org.uk. www.landmarktrust.org.uk.

WHILE YOU'RE HERE

Thanet has its own family-run pub chain, **Thorley Taverns**, offering good food and sometimes accommodation. It's worth looking out for its 18 traditional venues scattered along the coast. www.thorleytaverns.co.uk.

FURTHER INFORMATION

For more ideas on what to see and do, go to www.visitthanet.co.uk and www.visitkent.co.uk. Check opening times at all beer venues and attractions before you go.

PERFVGIVM MISERIS
A·D·MDCCCXLII

THE MARIS & OTTER

On the main road between Brighton and Hove, Harvey's has given this pub a new lease of life, with exposed brick walls and plenty of large tables to gather around. (If you're new to how beer is made, check out the diagram on the back wall). Upstairs are four suites named after varieties of kelp because of the brewer's link to the Sussex Kelp Project. Like downstairs, the bedrooms are modish, colourful and comfortable, with king-sized beds and enough room for a sofa. Little extras include Bluetooth Roberts speakers, and the kitchenette, with Nespresso machine, is a real bonus. You'll be spoilt in the bathroom, with a choice of walk-in shower and rolltop bath.
www.themarisandotter.co.uk.

THE MILLERS ARMS

Once the local boozer for workers at the mill that used to stand opposite, Shepherd Neame's 19th-century Millers Arms is in a pretty spot by the canal. It's relatively quiet here, a little away from the city centre, yet within easy walking distance of Canterbury's main attractions. Occupying the two upper floors above the pub, the dozen comfortable rooms are simply decorated and a mixture of shapes and sizes with doubles, twins and singles. Most offer pleasant views – some of the cathedral nearby. Service is friendly, informal and helpful, and the food downstairs, including breakfast, is a notch above regular pub fare to enjoy with well-kept Sheps draught beers.
www.millerscanterbury.co.uk.

MORE PUBS WITH ROOMS

BRIGHTON
The Southern Belle [33]
www.thesouthernbelle.co.uk.

CANTERBURY
Falstaff [34]
www.thefalstaffincanterbury.com.

FAVERSHAM
The Sun Inn [35]
www.sunfaversham.co.uk.

MARGATE
The George & Heart
www.georgeandheart.com.

BROADSTAIRS
The Royal Albion [36]
www.albionbroadstairs.co.uk.

[33]

[34]

[35]

[36]

SOUTH WEST

Opposite: Cotswolds.

BRISTOL

Brimming with brilliant breweries and charming neighbourhood pubs that are loyal to the local brews, all laced with a rebel spirit, the Bristol beer scene has a character all its own. While the city's attractions have been tackling their historic links with the slave trade, there's a positive vibe and lots of fun things to see and do.

CENTRAL & WATERSIDE

 ### WHERE TO DRINK

Good Chemistry Brewing took over the **Kings Head** [1], a short walk from Temple Meads Station, in 2022, rescuing one of Bristol's most interesting pubs. The peculiar interior, with its panelled 'Tramcar Bar', is Victorian, but there's evidence of a pub here since 1660. Seven keg lines pour mostly Good Chemistry's beers, as do the four handpumps. https://goodchemistrybrewing.co.uk/our-pubs-2.

The Cornubia [2] is another remarkable survivor, an 18th-century beery oasis with a large garden besieged by office blocks and building sites. Up to half a dozen cask ales are served. Also worth a visit, on a narrow cobbled lane with iron-edged kerbs nearby, is famous old ale house the **Seven Stars**. http://thecornubia.co.uk. https://7stars.co.uk.

Overlooking Bristol Harbour opposite Castle Park, the **Left Handed Giant Brewpub** at Finzels Reach is a modern taproom on the water's edge. The brewery occupies one side of the building while a lofty atrium fronts the bar, where you can choose from 20 LHG beers, including low-alcohol, on keg plus two on cask and a couple of guests besides. Tours are available. www.lhgbrewpub.com.

Across the river, into the old city, the **Strawberry Thief** is Bristol's premier Belgian beer specialist, with the likes of Palm, Brugse Zot, Westmalle Dubbel Tripel Karmeleit, Kasteel Rouge and Blanche de Bruxelles on tap alongside a couple of UK craft beers. And as you might expect, there's an impressive choice of bottled brews, too. https://strawberrythiefbar.com.

Brewpub **Zerodegrees** [3] is a spacious venue perched on a steep bank and split over two levels with an outdoor terrace that looks across the city. Four core beers are available on keg plus as many as half a dozen specials for the on-site brewer to test their ingenuity on anything from alcohol-free to milkshake pale. Over by College Green, the **Lime Kiln** is a rather more rustic beer house with a dozen lines shared equally between cask and keg, including a couple of locally brewed lagers.
www.zerodegrees.co.uk/restaurants/
    zerodegrees-microbrewery-bristol.
www.facebook.com/LimeKilnPub.

King Street is one of Bristol's big nighttime destinations, packed with bars, restaurants and an awful lot of beer. The **Famous Royal Navy Volunteer** is a bit of a mouthful if you have to ask the way, but it's worth it. No fewer than 30 lines take care of every imaginable style and there are always eight on cask. Next door, **Beer Emporium** is a large cellar bar with a long counter down one side and 24 keg taps offering a mixture of local craft and Belgian classics plus three cask ales on gravity at the far end. **Llandoger Trow**'s 28 keg lines include continental beers and British and American craft alongside up to five lines of cask – Holt's Mild was surprisingly on at last visit. And on the corner opposite the river, **King Street Brew**

House showcases at least a dozen of its own-brewed beers, cask and keg, in various styles.
www.facebook.com/VolleyBristol.
www.facebook.com/bristolbeer.
www.facebook.com/llandogerbristol.
www.kingstreetbrewhouse.co.uk.

On the other side of Queen Square, **Shakespeare Tavern** is a characterful 18th-century dockside inn currently operated by Greene King. The Suffolk brewer's Abbot Ale and IPA share the pumps with a few well-chosen local cask beers. It's next to the **Arnolfini** arts centre where Bristol Beer Factory runs a bright and busy café-bar that spills out onto the harbour in summer. A generous range of the company's beers are served on keg, and there's more across the water on Wapping Wharf at the **Junction**, a cheerful, pubbier venue with four cask ales alongside 14 keg taps pouring guests as well as its own beers. Perhaps the best place, though, to sample the Bristol Beer Factory range is the **BBF Tap Room** at the brewery itself, further out to the west, where five cask lines and eight keg showcase the latest brews.
www.greeneking.co.uk/pubs/city-of-bristol/
    shakespeare.
www.bristolbeerfactory.co.uk/pages/the-arnolfini.
www.bristolbeerfactory.co.uk/pages/junction.
www.bristolbeerfactory.co.uk/pages/the-bbf-
    tap-room.

It's hardly surprising that this multicultural, civic-minded city should be troubled by the key role played by previous Bristolians in the Atlantic slave trade, especially in the mid-1700s. There's no getting away from the fact that its profits underpin the city's institutions, fine Georgian architecture such as **Queen's Square**, and its museums and galleries. Step into the echoing atrium of **Bristol Museum & Art Gallery** and a display immediately acknowledges that this lofty Edwardian building was funded by the Wills family from its ill-gotten tobacco fortune and explains the decolonising work being done by today's museum team. As you go, you'll find labels have been updated to tell the back stories as well as descriptions of objects to give a clearer context to the assortment of rich global pickings gathered here, from Egyptian mummies to dinosaur skeletons and Old Masters. www.bristolmuseums.org.uk/bristol-museum-and-art-gallery.

The theme of slavery continues down the road at **The Georgian House**, the preserved home of plantation owner and slave trader John Pinney, and, on the harbourfront, **M Shed**, a multifaceted repository for Bristol's stories, is home to the defaced statue of the slave trader Edward Colston, famously dumped in the dock by protestors in 2020. The nearby Colston Hall reopened in 2023 following a major refurbishment with a new name, the **Bristol Beacon**. www.bristolmuseums.org.uk/georgian-house-museum.
www.bristolmuseums.org.uk/m-shed.
https://bristolbeacon.org.

But Bristol isn't crippled by the past. The waterfront is also where you'll find **We The Curious** [4], a fun, interactive quest for answers to questions about the world of science, the **Arnolfini**, which showcases visual and performance arts at the cutting edge, and **Create**, featuring an Ecohouse to promote ideas for greener living. (You'll find other sustainability projects around Bristol, the UK's first European Green Capital in 2015.) There's also a pleasant walk around the water, The Harbourside Loop, a footpath that only diverts inland where it's forced to by the fences around Brunel's **SS** Great Britain. While you must pay to even glimpse the famous transatlantic steamboat, a little further on, at the entrance to the North Entrance Lock, you can see a lesser-known work by Brunel for free, his **Swivel Bridge**. It's an important yet neglected piece of engineering, cared for by campaigning volunteers. www.wethecurious.org.
https://arnolfini.org.uk.
www.createbristol.org.
www.ssgreatbritain.org.
www.brunelsotherbridge.org.uk.

[5]

UNIVERSITY, KINGSDOWN & EAST BRISTOL

 ## WHERE TO DRINK

Popular with students, the **Robin Hood** has a bohemian vibe and a beer range that focuses on Bristol-brewed. Up to six pumps pour cask ale and there are 10 craft taps featuring Arbor, New Bristol and more. Further up the hill, former Dawkins house the **Green Man** is a compact wood-panelled local that's also a good place to find local brews with Moor often on cask and Lost & Grounded on keg.
www.facebook.com/robinhoodbristol.
www.facebook.com/thegreenmanbristol.

One of the city's unspoilt gems, **Highbury Vaults** has bags of character in all shades of brown. Don't be put off by the five handpumps in the tiny front bar all apparently serving Young's Ordinary (now known as Original) – they like a joke, here. The other bar has the full range, and a bar billiards table. For a completely different atmosphere simply cross the road to **Beerd**, a pizzeria-cum-craft beer bar with eight taps leading on Lost & Grounded's lagers and ales.
www.highburyvaults.co.uk.
https://beerdbristol.co.uk.

Another pair of contrasting venues are found on Cotham Hill. Bristol's **Brewhouse & Kitchen** has a full range of beers, cask and craft, brewed on the premises, plus guests, while **Coffee & Beer** is a small modern café that does what it says on the tin with up to five well-known names on the taps and an extensive range of cans. Keep heading north and you'll come to Good Chemistry Brewing's **Good Measure**, a former wine bar where you can drink your way through the Bristol brewer's range, on cask and keg, in a relaxed and friendly environment.
www.brewhouseandkitchen.com/venue/bristol.
www.coffeeandbeer.co.uk.
https://goodchemistrybrewing.co.uk/our-pubs-2.

Clinging to the side of the hill on the other side of Kingsdown, **Hillgrove Porter Stores** [5] has one of the city's best cask ale ranges, as well as some fascinating architectural features. Its 14 pumps are now joined by half a dozen craft taps including lagers from Lost & Grounded and Bristol Beer Factory. A few doors down, the **Hare on the Hill** is a traditional, green-tiled local with some well-chosen craft alongside the cask including from Kernel and Siren at last visit.
www.instagram.com/thehillgrove.
www.facebook.com/hareonthehillbristol.

At a lower altitude, **Basement Beer** [6] is just off busy Stokes Croft behind a heavily graffitied frontage where it brews and serves its modern beers from five taps, including an ice tea pale and a cocoa nib stout. Back on the main road, **Canteen** is a sprawling venue combining live music, imaginative vegan food and a good variety of Bristol-brewed beers on cask and keg.
www.basementbeer.co.uk.
www.canteenbristol.co.uk.

Cool and quirky, **New Bristol Brewery** also makes great beer, and you can test out the regulars and the new releases at its taproom on the brewhouse floor. The 15-line beer list is helpfully divided into categories such as 'crisp & bright', 'toasted & nutty', 'dark & roasted' and 'hoppy', of course.
www.newbristolbrewery.co.uk.

In East Bristol, the **Volunteer Tavern** is most famous for its beer garden, which is splendid, but take a look at those beers before you dash out there: 14 lines of craft and cask pouring Bristol's best, including lagers, pales, stouts and alcohol-free. The **Swan with Two Necks** has more of a punk ale house vibe and even more lines featuring, at last visit, a lot of Moor plus Deya and a Kernel saison.
https://sites.google.com/view/volunteertavern.
www.facebook.com/swan2necks.

Not to be missed, the **Wiper & True Taproom** at the Old Market is a wonderful place to drink beer, spacious and bright with views of the forest of gleaming vessels in the brewery, and a big foliage-decked garden for those al fresco moments. Sixteen lines pour every style you could wish for, and there are regular tastings and brewery tours.
https://wiperandtrue.com.

Kelly Sidgwick,
Good Chemistry Brewing

Good chemistry in brewing is a mixture of science, art and, because you sadly have to make a profit, business. Kelly Sidgwick brings the latter to the aptly named Good Chemistry Brewing, which she set up with partner and brewer Bob Cary in 2015.

'We wanted to do something creative with something tangible at the end of it,' she says. 'I'd run a business marketing elderflower drinks, so we thought we'd make a good combination. We're also called Good Chemistry because we believe beer brings everything together – science, arts, music, food and people.'

It soon became obvious that the firm should have its own pubs, and it opened the Good Measure in 2018 and the Kings Head four years later.

'There's a lot of competition among Bristol's brewers, but they all have their own spin on it. The city has some great taprooms, lots of different beer styles, great pubs and less traditional venues. One thing I think is pretty unique about Bristol is its support of local. It's not unusual to go into independent pubs and bars and see 80 per cent or more Bristol breweries on the bar.'

On Kelly's initiative, Good Chemistry's pubs host monthly She Drinks Beer events that encourage women to get together and share their appreciation of the drink.

'Coming into beer has been a steep learning curve for me,' she explains. 'That makes it all the more important I stand up and say to other women you can drink beer, you can get into brewing.'

https://goodchemistrybrewing.co.uk/
  she-drinks-beer/.

To the south, the **Left Handed Giant St Philips Taproom** [7] is more industrial, if you like to be in among the tanks of a working brewery. There are up to 11 lines pouring, including at least one cask, Saturday brewery tours and a busy programme of events.
https://lefthandedgiant.com/pages/tap-room.

In a triangle around the quaintly named Dings Park, you'll find Bristol Beer Factory's **Barley Mow**, a great little local with an impressive beer list including 18 draught lines; **Little Martha Brewing**, which sits under the railway arches and features well-chosen guests from the likes of Elusive, Moonwake and Burning Sky alongside its own; and **Moor Beer** [8]. The latter has a comfortable pubby taproom where you can spend your time well tasting this brewer's fusion of Californian and British styles, showcased on a dozen craft lines and at least one cask. Tours available.
www.bristolbeerfactory.co.uk/pages/the-barley-
  mow.
https://littlemarthabrewing.co.uk.
www.moorbeer.co.uk/brewery-tap.

It's worth noting Arbor and Good Chemistry open their doors twice a year to join Wiper & True, Left Handed Giant, Moor and Little Martha on the **East Bristol Brewery Trail**. Perhaps a good time to plan your visit!
www.eastbristolbrewerytrail.com.

[10]

WHILE YOU'RE HERE

The Lost & Grounded brewery is further out, beyond St Philips Marsh, but worth the trip to its taproom at weekends for a range of beers inspired by continental styles. You can do a brewery tour, too.
www.lostandgrounded.co.uk.

WHAT TO SEE & DO

The **Royal Fort Gardens**, on the University of Bristol's main campus, were designed by the 18th-century landscape architect Humphry Repton and are a place to picnic while studying some interesting **installations**, including *Follow Me*, a mirror labyrinth by the Danish artist Jeppe Hein. There are other prestige works around the university buildings, such as Luke Jerram's *Palm Temple* next to the School of Chemistry.
www.bristol.ac.uk.
https://public-art.bristol.ac.uk.

Opposite: **Clifton Suspension Bridge:** There are lots of places from which to admire Brunel's Clifton Suspension Bridge, a gorge-striding span bookended by two mighty sandstone towers. Sail beneath it, walk across it, or view it from the Clifton Observatory or Leigh Woods. The spectacle is even more beautiful when hot air balloons float over it during the annual August festival.

https://cliftonbridge.org.uk.

Several works by Bristol's most famous graffiti artist **Banksy** [10] can be seen around the city (there's a Banksy trail on the tourist board's website), including *The Mild Mild West* in Stokes Croft. But Banksy isn't the only show in town, as proved by the outdoor gallery around the Jamaica Street HQ of community activists **People's Republic of Stokes Croft**. Europe's biggest festival of street art and graffiti, **Upfest**, also takes place in Bristol.
https://prsc.org.uk.
www.upfest.co.uk.

Another important street mural can be found in St Paul's, the portrait of Jamaican-born Bristolian **Roy Hackett**, who led the successful Bristol Bus Boycott, protesting a ban on Black and Asian drivers and conductors, which led to the Race Relations Act of 1968. Hackett was also instrumental in setting up the **St Pauls Carnival**, a celebration of the local community's predominantly African-Caribbean heritage, which takes place in July.
Roy Hackett Mural, Byron Street, St Paul's, BS2 9NT.
www.stpaulscarnival.net.

FURTHER INFORMATION

For more ideas about what to see, go to https://visitbristol.co.uk. Check opening times at beer venues and attractions before you go.

BATH

Bath makes a fine beer destination with quirky pubs, craft beer bars and a modern brewery, all readily linked together via walkable city streets and riverside paths. The Georgian crescents and circuses and Roman Baths never fail to please, even if you have to share them with many other visitors.

 ## WHERE TO DRINK

Bath, unsurprisingly, has its fair share of historic pubs, and at the top of the list has to be the **Star Inn** [11] to the north of the city centre, away from the crowds. Four wood-panelled rooms are gathered around a central bar and many of the features date back to its origins in the 18th century – right down to the complimentary punch of snuff. Sitting there with your pint, or enjoying a game of shove ha'penny, you could be in another time. It's owned by **Abbey Ales**, the city's oldest brewer, and offers a good choice of cask beers from other brewers, including draught Bass on gravity. www.facebook.com/thestarinnbath.

For a totally different experience, two minutes further down the road Bradford on Avon-based **Kettlesmith Brewing** has a modest tap in a former shop where you can sample its eclectic range of craft beers from IPAs to saisons to a nice take on Kolsch. https://kettlesmithbrewing.com.

On your way back into town, drop into the **Bell Inn**, a buzzing cask-led music venue owned by the local community. Don't forget to check out the garden at the back. Or cross the river to a greener part of the city for a pint at the **Pulteney Arms** near the Holburne Museum, a pub you just know cares about its beer, serving some classic quaffable cask ales, among them Wye Valley HPA. https://thebellinnbath.co.uk/. www.thepulteneyarms.co.uk.

Bath Brew House is another must-stop for the beer lover. Founded in 2013, this modern brewpub has loads of space, including an extensive garden that hosts live music. Beers brewed on the premises are served on cask and keg and feature pale ales, lagers and continental styles, supplemented by a few well-chosen guests. The pub also runs regular brewery tours, tastings and brewing experience days when you can take part in the magical process. www.thebathbrewhouse.com.

Heading towards the centre, **Kingsmead Street Bottle** is a small shop selling craft beers in cans that also has a few well-selected brews on tap to drink on the premises, and, in fine weather, its tables spill out into this busy part of Bath. https://palmerstbottle.co.uk/kingsmead-street-bottle.

[11]

The Raven of Bath is a quirky freehouse over two floors with beers on keg and cask, including own-label brews made exclusively for the pub by Blindmans Brewery in the Mendip Hills. It's also home to the city's Storytelling Circle. www.theravenofbath.co.uk.

A few doors down there's another famous Bath pub, **The Salamander**, which is owned by Bath Ales, now part of Cornwall's St Austell Brewery, offering a range from both brewers on the pumps as well as being serious about its food. The bar is narrow and gets very busy, but there's more space upstairs. www.salamanderbath.co.uk.

The **Coeur de Lion** is an Abbey Ales pub with a sense of history about it, and perhaps the best place to try the brewer's flagship Bellringer on cask. It's tiny and the staircase up to the overspill bar treacherous, but it all adds to the character. www.facebook.com/Coeurdelionbath.

Bath's big player in the craft beer world, **Electric Bear Brewing**, has a taproom at Newbridge, to the west. The best way to get there is simply to walk along the river path from the city centre. You'll find a varying range of beers on 10 lines, mostly IPAs, which you can enjoy surrounded by brewing vessels or out in the yard. Brewery tours are available, too. https://electricbearbrewing.com.

On the way back, be sure to stop off at the **Royal Oak** on the Lower Bristol Road. There are other Royal Oaks around – this is the one with a brewery, Ralph's Ruin, and a diverse array of beers on cask and keg, including interesting Belgians. It's also a live music venue. Grungy, but in a good way. www.theroyaloakbath.co.uk.

A couple of other places to note. Above Bath Spa Station, looking out onto the tracks, Bath Ales has an upstairs sports bar called **Ludo** with craft beers from Harbour Brewery on tap. Meanwhile, Guinness fans will point you to **Flan O'Briens** for the best pint of the black stuff in town. https://ludobath.co.uk. www.facebook.com/flan.obriensbath.

Do you need a guide to Bath? It is surely one of the most written-about destinations on these isles – and in multiple languages. In fact, you could just follow the crowds around this amphitheatre of limestone crafted by the Romans and Georgians with such skill that is has been declared a UNESCO World Heritage Site. You might as well be swept along the mellow yellow streets with the pack to the major sights – John Wood the Elder's **Circus** and **Queen's Square**, John Wood the Younger's **Royal Crescent** [12], and Robert Adam's **Pulteney Bridge**, which was probably inspired by the rather famous spans in Florence and Venice.

But get a head start at the **Roman Baths**. It pays to get here early and join the walk-in queue or book a timed ticket for when it opens, otherwise be prepared for a stroll around its interior to slow to a shuffle. (You can while away time as you wait in the queue by studying the angels climbing up and down ladders on the West front of **Bath Abbey** [13] – pop inside later to see the fan-vaulted ceilings.) Either way, the Baths are unmissable, displaying so many astonishing remnants of the complex that the Romans built around a thermal spring and the embellishments made by the Georgians when they turned this city into the 18th century's top spa resort.

Most impressive is the Great Bath [14], which Roman emperors and governors watch over from the balcony above. And the curators have played a clever trick illuminating the now-plain fragments of the Temple pediment, including the head of a scary Gorgon, to show its original rainbow colours. An audio tour included in the entrance fee is not only useful for understanding what you're looking at but comes with the entertaining options of hearing the reactions of the writer Bill Bryson (he's not fond of the gilt bronze head of Minerva) and a children's commentary by the writer Michael Rosen that will amuse adults, too. You can sip the waters, in the Georgian style, at the end in the Pump Room and even bathe in them (for a price) along the road at the **Thermae Bath Spa**. www.romanbaths.co.uk. www.bathabbey.org. www.thermaebathspa.com.

[12]

[13]

Rugby fans can glimpse the stadium from the railings at the end of William Street. Further along, on Sydney Place, note the plaque at number 4 that explains this was Jane Austen's home for four years, before entering at the **Holburne Museum**. What's here? A variety of art and artefacts gathered around Europe by the globetrotting baronet of the same name (using money earned in the plantations of the Caribbean, which is acknowledged). There are Gainsboroughs, Turners, Breughels and more, but don't miss the detailed miniatures and mourning jewellery. The building looks like a house but was always just the entrance to **Sydney Pleasure Gardens**, which is still a lovely place, with some spectacular trees, to wander around. It leads to the Kennet & Avon Canal for peaceful walks along the towpath. www.holburne.org/.

If you do go to the Circus, visit the **Bath Assembly Rooms**, too. Formerly the Fashion Museum (which was moving to the Old Post Office at time of writing), something new is coming here and it will fully reopen in 2026. Meanwhile, you can still book tours of the building. On the corner, the **Museum of East Asian Art** is filled with objects and furnishings from the region spanning 7,000 years, from tiny, intricate scenes sculpted in jade to large, lacquered cabinets. www.nationaltrust.org.uk. https://meaa.org.uk.

Between the Circus and the Royal Crescent, dip down **Margaret's Buildings**, an attractive pedestrianised Georgian terrace where cafés spill out onto the flagstones on warm days. Browse the heaving shelves of **Bath Old Books** (there's another good bookshop, **Topping & Co**, in the old Friends Meeting House on York Street), the exquisite interiors of pricey homewares shop **Berdoulat**, and work by living British and Irish artists on display in the **Red Rag Gallery**. https://uk.bookshop.org/shop/BathOldBooks. www.toppingbooks.co.uk. https://berdoulat.co.uk. www.bathartgallery.co.uk.

FURTHER INFORMATION

For more ideas on what to see and do, go to https://visitbath.co.uk. Check opening times at beer venues and attractions before you go.

CHELTENHAM

Cheltenham seems an unlikely beer destination, better known for its racecourse and a generally posh air. But it has developed a rich and varied pub and bar scene – and, in Deya, boasts one of the country's leading craft brewers. It's also in the premier league of Britain's historic spa destinations, with the best Regency landscape to stroll about in the country.

 ## WHERE TO DRINK

Deya Brewing [15] is on an industrial estate close to Cheltenham Spa station. If you're in the centre of town it's a mile or so walk, but there's a pretty path avoiding the traffic designed for commuters but equally good for beer tourists. Starting at the Jessops Avenue playground, it follows the River Chelt before detouring to the station. The brewery has a large on-site taproom and yard filled with long trestle tables beer hall-style and a bar at the end with at least 20 keg lines plus a handpull serving one of Deya's occasional cask ales. Familiar brews such as Steady Rolling Man and Magazine Cover are always available, along with specials and new releases, mostly, but not exclusively, pales and IPAs, including a couple of guests. Around the corner there's the smaller and cosier **Deya Mixed Fermentation Taproom**, where you can do your drinking among the oak barrels.
www.deyabrewing.com.

Arguably the town's best-known pub among beer lovers, **Sandford Park Ale House** is a former Campaign for Real Ale pub of the year and it never disappoints. Ten cask lines and 16 keg offer a huge choice of beers, all served with care by knowledgeable staff, and there's plenty of space to drink them, including a large garden and an overspill bar upstairs.
https://sandfordparkalehouse.co.uk.

Nearby there's a cluster of places worth a look around the high street. **The Strand** is a spacious upmarket bar-restaurant with a 20-strong beer list taking in some of the best UK brewers, including four on cask. Across the road, **The Swan** is more traditional and offers craft beers from local brewers, including Stroud and Tiley's, to wash down some creative Indian street food, while **The Vine** is a classic, green-tiled Victorian corner house with local brews on cask and keg.
www.facebook.com/TheStrandGL50.
www.theswancheltenham.co.uk.
www.thevinecheltenham.co.uk.

Around the corner, painted bright pink, you'll spot **Planet Caravan**, a small, friendly neighbourhood bar on two floors with a generous 16 craft lines in styles to suit all tastes. It has a canning machine, too, so you can take away a draught beer – but it's the sort of place where you'll want to linger.
www.planetcaravancheltenham.com.

[15]

The Montpellier area of town has a pretty, and very popular, local called **The Beehive**, which is famed for its local cask ales, while in the backstreets further south the **Jolly Brewmaster** features a broader range on the pumps and has a fabulous quirky beer garden in which to enjoy your selection.
https://thebeehivemontpellier.com.
Jolly Brewmaster, 39 Painswick Road, GL50 2EZ.

On the main road south of the A40, among the shops, **Bath Road Beers** is good for the connoisseur and beginner alike, with an extensive range of bottles and cans, five draught lines, including one for cask, and a growler station for takeaways, plus a tasting room at the back that hosts tastings and food pairings. If you're dazzled by the array, knowledgeable staff will help you choose.
https://bathrdbeers.co.uk.

Brewhouse & Kitchen has a branch in the Brewery Quarter, on the site of the old Flowers Brewery, which closed in 1998. It's nice that someone's still brewing here, and it could be you if you sign up for one of the brew days offered on the menu of beer experiences. Or just enjoy sampling the wide range on the bar.
www.brewhouseandkitchen.com/venue/
    cheltenham.

Kemble Brewery Inn no longer brews but it's a very attractive local pub hidden away off the road to the racecourse. Six cask beers are served and there's a special welcome for racegoers and students of the turf.
www.facebook.com/thekemble.

WHAT TO SEE & DO

When, in the 18th century, William Mason tapped into a natural spring in a meadow near where Cheltenham Ladies College now stands, his brainwave to capitalise on a trend among the wealthy of the day, to take curative waters, put Cheltenham in the game to join the premier league of England's fashionable spa towns and become Britain's foremost showcase of Regency architecture.

Wherever you roam in Cheltenham there's something from that period to inspect. Even on the main shopping drag you'll glimpse orderly terraces of white stucco along the side streets that part the global chain stores. The hotspots are the **Pittville Pump Room and Park**, the **Royal Crescent**, and the genteel residences of **Tivoli and The Suffolks**. Here the architects of the day gave free rein to their interpretations of Classical Greek and Roman design, embellishing buildings with wrought-iron railings, porticos and verandas, ornate pillars, friezes, and balustrades. The whole effect comes together particularly well in **Montpellier**, a neighbourhood that developed around a spa in the 19th century with the old pump room and domed Rotunda ballroom (now The Ivy restaurant), private pleasure gardens (now a public park), and a terrace of shops the fronts of which are supported by 32 graceful female figures, or caryatids, inspired by the Acropolis of Athens. https://pittvillepumproom.org.uk.

This is not the only attractive shopping parade. Follow the road north-east to **The Promenade**, an avenue of horse chestnut and elm trees bordered by civic gardens and buildings and smart shops and cafés, with another manicured green space, Imperial Gardens, and the ostentatious **Neptune Fountain** [16], based on the Trevi in Rome.

The Promenade is also the location of The Hare and the Minotaur, two giant bronze creatures sculpted by Sophie Ryder, who watch the world go by from their bench outside the House of Fraser department store. Just around the corner on Clarence Street is some more contemporary artistic relief, a bronze by Barbara Hepworth in the form of a trio of semi-circular abstract forms, titled Themes and Variations, on the old HQ of the Cheltenham & Gloucester Building Society. And if you dip down Grosvenor Place South, you'll find an amusing sequence of tiled panels depicting the chaos that ensued when an elephant escaped from a circus parade in the town in the 1930s.

[17]

You could spend most of your time wandering around Cheltenham gawking at all this, but there are specific sights to see. The **house where Gustav Holst was born** [17] in 1874 is now a museum dedicated to the composer and the times in which he lived. **The Wilson** has a fine collection of Arts and Crafts furniture and the archive of Edward Wilson, who accompanied Scott on his fateful final trip to the Antarctic. But these days, if you're not here for a Regency overload, you're likely to be present for a festival. Jazz, science, music literature and horse racing – Cheltenham excels in these showpiece events.
https://holstvictorianhouse.org.uk.
www.cheltenhammuseum.org.uk.
www.cheltenhamfestivals.com.
www.thejockeyclub.co.uk.

WHILE YOU'RE HERE

The **Plough Inn** is within a furlong or three of Cheltenham Racecourse, by the church in the quaint village of Prestbury. The Grade II-listed thatched main building has two small bar areas, one with an open fire, and an unusual bar arrangement with two serving hatches. The attraction for the beer enthusiast is that the local cask ales are served on gravity. That's not easy to get right, but the Plough manages it. And it has a splendid garden, too.
www.theploughprestbury.co.uk.

FURTHER INFORMATION

For more ideas on what to see and do, go to www.visitcheltenham.com. Check opening times at beer venues and attractions before you go.

THE COTSWOLDS

There are many beautiful pubs in the Cotswolds, and we've chosen a couple of areas where we think the beer tourist will find most satisfaction. The mix includes one of the area's deservedly famous honeypots and a cluster in the south you might not know but should explore.

 ## WHERE TO DRINK

Donnington Brewery [19] has 19 pubs in these hills, and perhaps the most convenient place to sample its beers, close by in Stow-on-the-Wold, is the **Queen's Head**. On the town square, this 17th-century inn has bags of character and history. It has served Donnington's ales for more than a hundred years and today you'll still find the full range on the pumps.
https://donnington-brewery.com/pages/
  the-queens-head-stow-on-the-wold.

While in Stow, make sure you pop into **Off the Square** (you can guess where it is), a surprising little bar and bottle shop focused on modern beers and featuring the likes of Deya, Burnt Mill and Brew York on tap. It also hosts a range of events including tutored tastings.
https://offthesquarestow.co.uk.

Across to the south-western side of the Cotswolds, the **Old Spot Inn** at Dursley is one of the country's most celebrated beer pubs. At least half a dozen cask ales feature on the bar alongside a few craft taps, served in suitably rustic surroundings.
http://oldspotinn.co.uk.

In neighbouring Nailsworth, the **Village Inn** [18] is an eccentric, sprawling brewpub that's home to Keep Brewing and a beer or two you may not find anywhere else. **The Fresh Standard Brew Co**, found at the back of an industrial estate on the edge of town, offers a range of modern styles straight from the cask in its tiny tap room.
www.villageinn-nailsworth.co.uk.
https://thefreshstandard.co.uk.

[20]

Stroud Brewery [20] is a must-visit. You'll find its all-organic beers in many of the local pubs but its headquarters on the A419 serves the full range on cask and keg in its large taproom and yard beside the canal. There are two more floors of events spaces up the stairs and there's always something going on. The best way to approach it is along the towpath. https://stroudbrewery.co.uk.

The nearby **Ship Inn** is a more traditional setting with a well-chosen selection on half a dozen handpumps plus a few interesting options on the keg taps, while in the centre of Stroud the **Ale House** is an unusual set-up in a grand building that was headquarters to the Poor Law Guardians. It now has a striking bar pouring a fine selection of 11 cask beers from across the country. www.theshipstroud.co.uk. www.thealehousestroud.com.

👓 WHAT TO SEE & DO

Sheep once filled the market square at **Stow-on-the-Wold**, now cars flock to this vast space lined with buildings cut from the local honey-coloured limestone. Their occupants come to wander about this impossibly pretty town, which made its fortune from the wool of the beasts once herded – and counted – through tight alleyways, or 'tures', into the square. Unpack more history – from the Iron Age to the English Civil War – on a **Town Walk**.

The rise of Stow's wool market from the 15th century shapes the town landscape, today a place to browse boutiques, antiques shops and galleries (the **Fosse Gallery** is the most interesting). Be sure to see the north porch of the 12th-century church of **St Edward's**, out of which appear to grow the two yew trees flanking its arched wooden

door. Locals will tell you it's the inspiration for JRR Tolkien's Doors of Durin in *The Lord of the Rings*.
www.stowcivicsociety.co.uk/stow-town-walks.
https://fossegallery.com.
www.scats.org.uk.

You'll find fragments of history to piece together in **Dursley** at the local **Heritage Centre**, and the **Georgian Market House** provides a pretty space for the regular farmers market. But this town's strength is as a Walkers Are Welcome destination, with paths to Stinchcombe Hill and the Tyndale monument for sublime views of the Cotswold escarpment and the Severn Vale.
www.dursleyhc.org.uk.
www.facebook.com/dursleyfarmersmarket.
https://dursleywelcomeswalkers.org.uk.

Nailsworth has a similarly beautiful setting, cradled by woods, and its streets are gaining a serious reputation for independent shops such as **Cotswold Craftsmen Gallery** and foodie outlets of the calibre of **William's** food hall. Nailsworth is also home to **Forest Green Rovers**, a football club renowned for its vegan principles. The town also claims the largest number of working watermills per square mile in Britain. On open days you can see a waterwheel and weaving looms in action at **Dunkirk Mill Museum** and **Giggs Mill Weaving Shed**.
www.cotswoldcraftsmengallery.co.uk.
www.williamsfoodhall.co.uk.
www.fgr.co.uk.
www.stroudtextiletrust.org.uk.

Less green wellies, more tie-dye, **Stroud** has cultivated a bohemian look amid its hilly streets, with more than its fair share of shops selling vintage clothes, crystals and crafts, and veggie, organic and plant-based eateries, plus one of the country's biggest farmers markets. **The Shambles** has been trading since medieval times amid a cluster of buildings that, over the centuries, have been used as a school, pub, town hall and jail.
https://fresh-n-local.co.uk/trader/stroud.
www.shamblesmarketstroud.co.uk.

FURTHER INFORMATION
For more ideas on what to see and do, go to www.stowinfo.co.uk, www.dursleytowncouncil.gov.uk, www.nailsworthtowncouncil.gov.uk and www.stroudtown.gov.uk. Check opening times at beer venues and attractions before you go.

The scenic route: Follow the B4066 between Stroud and Dursley for superb views from Coaley Peak and Selsley Common over the Malvern Hills and River Severn to the Black Mountains in Wales. Selsley Common is protected because of its limestone grasslands, which are laced with wildflowers including the rare fly-bee hybrid orchid. At Coaley Peak, a hump in the grass reveals the presence of a Neolithic burial mound.
www.nationaltrail.co.uk/en_GB/attraction/selsley-common/

DONNINGTON BREWERY
Owned by the family behind the Arkell's Brewery in Swindon, Donnington Brewery occupies an old mill in an idyllic lakeside spot just outside Stow-on-the-Wold. It's hard to find down a steep, narrow track, but worth seeking out – especially now it's offering tours. There's nothing quite like it. The brewery is close to what it must have been like when Richard Arkell started making beer in the former mill building in 1865, and the mill wheel powered by a stream still drives some of the machinery. You can stay here, too, at the two self-catering holiday homes on site.
https://donnington-brewery.com.

CORNWALL

Since national brewers hesitated over trading in a remote part of the country where sales are seasonally distorted, Cornwall developed its own beer culture based around local brewers such as St Austell, Skinners and Sharps – now joined by craft aristocracy in Verdant and others. While you're here, you'll discover tin wasn't the only natural resource mined locally and a drift of exotic gardens nurtured by the mild climate.

ST AUSTELL & CHARLESTOWN

WHERE TO DRINK

Wherever you go in Cornwall, you'll find a **St Austell Brewery** pub and a couple of beers on the handpumps that you'll recognise. Alongside Tribute and Proper Job, though, you may discover a couple of local favourites in the family brewer's home county. In particular, look out for Hicks Special Draught, or HSD – a distinctive strong ale named after the company's founder.

But only a visit to the brewery itself can give you an idea of the surprising variety of beers that St Austell now brews. **Hicks Bar** [21], which is open to visitors during the day, showcases an extensive range of styles and experimental brews on cask and keg. It's always changing and there's always something interesting to try that you won't find anywhere else. How about a Mango Baobab Cornish Sea Salt Sour? Or a Pina-Co Lager? Perhaps that's pushing the boat out too far, but you're bound to try something you like, and you're bound to be impressed by the versatility and creativity 'traditional' brewers can demonstrate.

Next door, small mountains of bottles display St Austell's more successful experiments, among them Big Job, the 7.2% abv version of Proper Job; the latest vintage of Black Square Imperial Stout; special collaborations and a Cornish take on Belgian beers. While you're here, get a sense of St Austell's long history by browsing the mini-exhibition, or book your ticket for a brewery tour or tutored tasting. https://staustellbrewery.co.uk. https://hicksbar.co.uk.

While you're in town, nip over to the picturesque Georgian harbour of Charlestown where at St Austell's **Rashleigh Arms** you'll find up to nine handpumps pouring, usually including a guest ale or two. In the summer of 2024, the **Harbour Beer House** opened down the road. A collaboration between St Austell and Harbour Brewing, it features exclusive craft beers and limited edition specials on cask.
https://rashleigharms.co.uk.
Harbour Beer House, Harbour Front, Charlestown Road, PL25 3NJ.

HE HICKS
MUSEUM & BREWERY
Specials

[22]

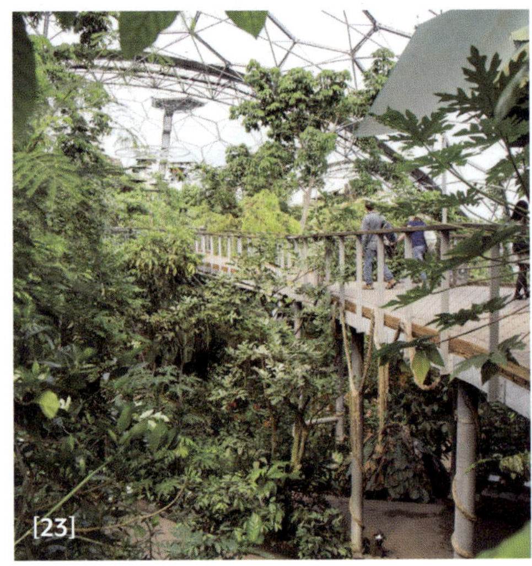

[23]

WHILE YOU'RE HERE

Across the bay, in the tiny fishing village of Polkerris, there's a great pub called the **Rashleigh Inn** [22] (confusingly, but the Rashleighs were local bigwigs). Not content to trade on spectacular harbourside views, this Rashleigh sources some hard-to-find Cornish beers, serving a carefully chosen range on tap and handpump. https://therashleighinn.co.uk.

WHAT TO SEE & DO

From the size of St Austell and some of the architecture it's obvious that this was once a wealthy community, a fortune made from 'white gold', aka china clay, discovered in the area in the 18th century. Little is mined here today – production has moved to Brazil – but the pyramids of waste product can still be seen on the skyline, hubristically nicknamed the Cornish Alps, and there are even trails on the tourist board's website that connect them on foot, bike or horseback. **Wheal Martyn Clay Works**, just outside town, explains the history in an authentic setting as well as offering a view of a modern working pit. www.claytrails.co.uk. www.wheal-martyn.com.

The best of St Austell can be found at its edges, including the **Eden Project** [23], where huge geodesic domes in a former clay pit hothouse plants normally found in rainforest and Mediterranean climates. The south coast of Cornwall is blessed with botanical gardens (see Cornish Gardens, page 209), which here includes **Pinetum Gardens**, where there's a giant redwood among the conifers and different plots to investigate, planted according to setting or style, such as the woodland and Japanese gardens. You can stay here, too. www.edenproject.com. https://pinetumgardens.com.

Clay – as well as tin and copper, also mined in these parts – led to the creation of **Charlestown**, which was just another fishing cove called Porthmeur until, in 1791, Charles Rashleigh transformed it into a harbour big enough for ships to drop anchor and named it for himself. Today, it is such a remarkably well-preserved Georgian harbour that it bears the UNESCO stamp. Its deep dock is encircled by wharves, warehouses, stone cottages, a couple of inns, and the remains of a gun battery on the cliff above. The masts of two resident replica tall ships, *Anny* and *Kajsamoor*, poke out above the harbour walls – you can usually go aboard – and the **Shipwreck Treasure Museum** on the quay has thousands of finds from wrecks to pick through. Though Charlestown is peppered with bars, cafés and craft shops and usually overwhelmed by visitors or film and TV crews (we are in the BBC's *Poldark* country), there's a reassuring sense that daily life goes on, too. https://charlestownharbour.com. https://shipwreckcharlestown.co.uk.

[24]

PENRYN & FALMOUTH

🍺 WHERE TO DRINK

Penryn is the home of **Verdant Brewing Co.** [24], one of the UK's most highly regarded craft brewers. Its spacious taproom, with seating indoors and out, has a glass wall at the back through which you can watch the brewery at work. Verdant on cask is rare outside Cornwall, but here you'll have three or four ales to choose from plus another 16 on keg featuring its signature deep-flavoured IPAs and new releases. Neapolitan-style pizzas are also served, and there's a busy programme of events including tastings. And it's not the only brewery on the

Kernick Industrial Estate. You might be lucky to find **Lacuna Brewing Co.**'s doors are open and try beers brewed with the environment in mind. https://verdantbrewing.co. www.lacunabrewing.com.

The town merges with Falmouth, one of Cornwall's premier tourist destinations with some interesting places to drink. For one thing, there's the legendary 17th-century **Seven Stars**, which has been in the same family for seven generations and has lost none of its character, steeped in the tales of the many salty seadogs who've passed through. Local ales on the pumps are supplemented by the beer the pub is perhaps most famous for – draught Bass served straight from the cask. www.thesevenstarsfalmouth.com.

Is it a pub? Is it a bookshop? In fact, **Beerwolf Books** [25] is both. Walk up a flight of stairs off Falmouth's main shopping street and you'll find yourself in the middle of a well-stocked library. Turn around, and you'll see the bar, with a half-dozen cask pumps and a good selection of craft beers on the taps. Plenty of seating, too, for you to drink and read. www.facebook.com/BeerwolfBooks.

Verdant Seafood Bar, the brewer's outpost in town, is a cosy nook tucked away on a side street running down to the harbour. Eight lines of craft keg are typically pouring alongside a constantly changing menu of imaginative small plates of fresh seafood. https://verdantbrewing.co/pages/seafoodbar.

The Front is the town's premier ale house, set on the quayside with a solid choice of beers on the pumps and, if the mood takes you, Breton dancing on Mondays, while the **Moth & The Moon** is a modern local a little away from the hubbub featuring Cornish brews on keg and cask. The Front, Custom House Quay, TR11 3JT www.facebook.com/mothandmoonfalmouth.

👓 WHAT TO SEE & DO

Penryn is another old community that has largely maintained the integrity of its historic landscape. Once a fish market, now its streets are fast becoming an artists' enclave, buoyed by the presence of a campus of the University of Falmouth, which majors in the creative arts. There are galleries at every other step along **Lower Market Street and Broad Street**, as well as vegan-friendly coffee shops and delis. More craftspeople work out of studios in repurposed buildings on **Commercial Road**, on the banks of the Penryn River, including the modern complex **Jubilee Wharf**. Should you feel the urge to sign up for a quick workshop on pottery or printmaking, you'll find it here. www.jubileewharf.co.uk.

Falmouth is also young and creative at heart, the students from the town's campus keeping the bars and cafés busy around The Moor (the town square) and along Market, Church and Arwenack streets, a thread of cobbles that runs parallel to the banks of the Fal Estuary. There are plenty of galleries to browse as well as exhibitions in **The Poly arts centre**, which also has a shop that's a cut above the usual, selling work curated by The Guild, an organisation founded in 2015 to champion local artists and makers. Its founders, The Cornwall Polytechnic Society, began nurturing creativity in this Cornish port in the 1830s, not long

CORNISH GARDENS
Blessed with the mild winters and warm
summers brought by the Gulf Stream,
Cornwall's south coast blooms with gardens.
They come in all sorts of guises – arranged
around manor houses, hidden in woodland,
even embracing the sea, and what grows in
them reveals tales about Victorian plant
hunters and the hard work of devoted
gardeners. Between St Austell and Falmouth
are some fine specimens, including **The Lost
Gardens of Heligan**, which was hidden in
undergrowth until the 1990s, and the National
Trust's **Trelissick House and Garden**, where
azaleas and rhododendrons sun themselves
by the Fal Estuary.
www.heligan.com.
www.nationaltrust.org.uk/visit/cornwall/
    trelissick.

[26]

after Turner visited and caught the special quality of the local natural light on canvas. It was a challenge eagerly taken up by subsequent generations and some of their work is on display at the **Falmouth Art Gallery**, in the Passmore Edwards building on The Moor. The gallery also has an extensive collection of automata.
https://thepoly.org.
www.falmouthartgallery.com.

There's maritime history to explore here, too, at this once strategic post at the entrance to the **Carrick Roads**, the other name for the Fal Estuary, from which waterways wind inland to the Cornish capital, Truro. **Pendennis** and **St Mawes Castles** [28] were built to guard them on the order of Henry VIII, who was fearful of European invaders. You can visit both in one day if you hop on the St Mawes ferry. This being the third largest natural harbour in the world, it's only fitting that it's the site of the **National Maritime Museum Cornwall** [26].
www.english-heritage.org.uk/visit/places/
    pendennis-castle/.
www.falriver.co.uk.
https://nmmc.co.uk.

WHILE YOU'RE HERE

On the north coast at Rock, **Sharp's Brewery** offers tutored tastings on site, so you can find out what they do aside from Doom Bar. Further to the west, at Trevaunance Cove, **Driftwood Spars** is a quirky pub with rooms and its own brewery where you might lend a hand with the brewing. **Skinner's Brewery**, enjoying a fresh lease of life after being rescued by local entrepreneurs, showcases a growing range of beers at its large taproom and yard on the outskirts of Truro, while in nearby Threemilestone, **Mason Brewing** opened its doors

in 2024. North of Penryn, **Dynamite Valley Brewing Co.** hosts its beer café at Porsanooth on Saturdays. And you must try to fit in the **Blue Anchor** [27] at Helston, home of **Spingo Ales** and one of the oldest surviving brewpubs in the country.
www.sharpsbrewery.co.uk.
https://driftwoodspars.co.uk.
https://skinnersbrewery.com.
www.dynamitevalley.com.
www.spingoales.com.

Sculptures by Barbara Hepworth are on view at her former studio in St Ives [29], part of Tate St Ives. And the backdrop of the ocean adds drama to performances that take place on the cliffside stage at the Minack Theatre.
www.tate.org.uk.
https://minack.com.

FURTHER INFORMATION

For more ideas about what to do and see, go to www.visitcornwall.com. Check opening times at beer venues and attractions before you go.

[27]

[28]

[29]

RASHLEIGH ARMS

At the centre of the Georgian fishing village of Charlestown, St Austell Brewery's flagship Rashleigh Arms was, in the 19th century, the building where they dried the china clay. Frequently visited by film crews, it boasts the only Grade I-listed cobbled car park in the country. Now it's a spacious modern inn with 18 comfortable, generously furnished rooms, split between the upstairs floors of the main pub and the annex at the bottom of the car park. Deluxe rooms have a view of the harbour and the sea beyond. It's a well-run operation with high levels of service offering a superior and varied locally sourced pub menu including, of course, hearty breakfasts and the freshest fish, alongside a full range of St Austell ales.

https://rashleigharms.co.uk.

DONNINGTON BREWERY

You can sup and stay at Donnington Brewery, which offers two self-catering options at its picturesque location. Old Brewery House is adjacent to the brewery with views across the private lake and gardens. It has four simply yet tastefully furnished en-suite bedrooms, which can be configured as doubles or twins. Sleeping eight to 10 people, this is one for a group of family or friends, with two reception rooms to relax in. A little away from the brewery, secluded in its own grounds for more privacy, Waterhead Barn is a bright, spacious modern conversion where care has been taken to retain much of the original structure. There are six en-suite bedrooms, sleeping up to 13, so even more can join you. There is a large open-plan living and dining room, and, adding a homely touch, an Aga in the kitchen. On sunny days you're sure to make use of the courtyard terrace. Even better, they can be booked together for larger groups, and up to two medium-size dogs are welcome at each.

www.donnington-brewery.com/pages/
our-luxury-cotswold-stays.

MORE PLACES TO STAY

BRISTOL
Hort's Townhouse
www.hortstownhousebristol.co.uk.

The Channings
www.greenekinginns.co.uk/hotels/avon/
channings-hotel.

The Wellington
thewellingtonbristol.co.uk.

BATH
Broad Street Townhouse
https://butcombe.com/broad-street-townhouse-
bath.

The Bear Inn
www.bearinnbath.com.

The Saracens Head [30]
www.greeneking.co.uk/pubs/somerset/
saracens-head.

CHELTENHAM
The George Hotel [31]
www.georgehotelcheltenham.com.

No.38 The Park
www.no38thepark.com.

STOW-ON-THE-WOLD
The Porch House
www.porch-house.co.uk.

The Stag at Stow
https://thestagatstow.com.

The Kings Arms
www.kingsarmsstow.co.uk.

The Old Stocks Inn
www.oldstocksinn.com.

Sheep on Sheep Street
www.thesheepstow.co.uk.

The Bell
www.thebellatstow.com.

NAILSWORTH
Egypt Mill
www.egyptmill.com.

ST AUSTELL
The White Hart Hotel
https://whitehartstaustell.co.uk.

FALMOUTH
Star & Garter
www.starandgarterfalmouth.co.uk.

Chain Locker
https://chainlockerfalmouth.co.uk.

The Thirsty Scholar
www.thethirstyscholar.co.uk.

7

SCOTLAND

Opposite: Anthony Gormley statue outside the Gallery of Modern Art, Edinburgh.

EDINBURGH

Edinburgh has long been the best place in Scotland to drink good beer, and the craft revolution has taken it to a new level, especially when you add in a trip to Leith. It's also filled with many of the nation's great cultural institutions, including a fine castle and the biggest arts festival in the world.

OLD TOWN

 ### WHERE TO DRINK

Salt Horse is a wonderful little beer café, a bright and modern, quiet and cosy haven from the crowds thronging the Royal Mile only a few minutes' walk away. Choose from 14 keg lines and around 250 bottles and cans brewed at home and abroad. www.salthorse.beer.

Just as friendly and welcoming, **Sandy Bell's** is a traditional local behind the National Museum of Scotland, famous for folk sessions you can accompany with a well-kept pint from Scottish brewers such as Harviestoun and Dark Island. https://sandybells.com.

Cold Town House on the corner of Grassmarket is different again. Spread across three floors in the shell of an old church, the home of Cold Town Brewery has plentiful taps serving beers brewed freshly on the premises plus guests. Its biggest attraction, though, is a roof garden with a view of the castle. https://coldtownhouse.co.uk.

On the Fountainbridge side of the city centre, **Innis & Gunn** [1] has a typically spacious tap offering a diverse selection of brews, and the **Hanging Bat** is an independent craft beer bar showcasing local brewers on cask and keg, while the **Blue Blazer** is more of a traditional ale house with a reputation for well-kept beer. www.innisandgunn.com. www.thehangingbat.com. https://kilderkingroup.co.uk/the-blue-blazer.

Heading south towards the Meadows, **Cloisters Bar** is a comfortable retreat set within the walls of a former parsonage. Its nine handpumps and 10 keg taps feature mostly Scottish brewers, including Swannay, Newbarns and Black Isle. **The Golf Tavern** is an atmospheric spot for a pint, dating back to 1456, and looks over Bruntsfield Links, one of the world's oldest golf courses. www.cloistersbar.com. https://golftavern.co.uk.

And if you fancy treating yourself, **Timberyard** is a restaurant with an amazing beer list led by Belgian lambics and mixed fermentation brews. www.timberyard.co/#first.

[2]

WHILE YOU'RE HERE

Hop off the train at Haymarket and nip into **Ryrie's Bar** next to the station. Family-owned and beautifully restored, this Edwardian gem offers a broad selection of cask beer from its seven pumps plus a couple of craft options. And a little way down the road you'll find the **Wee Vault** [2], a narrow shop converted into a tasting room for Edinburgh's sour beer specialist Vault City. The range punches above its weight with no fewer than 24 taps pouring a taste-bud tingling array of flavours. www.ryries.bar.
https://vaultcity.co.uk/pages/the-wee-vault-
   tasting-room-and-bottleshop.

Bellfield Brewery, which makes all gluten-free beers, has a fine taproom and yard to the north of Holyrood Park. A dozen keg lines serve mostly its own brews, which cover a variety of styles including IPAs, lagers, saisons and Scottish classics, including a couple on cask. On the way, **Jeremiah's Taproom** has 20 lines showcasing craft beers from around the UK. Or, a bit further out on Easter Road, **Old Eastway Tap**, operated by Cross Borders Brewing, features collaboration brews among craft offers from 16 keg taps and four cask lines. www.bellfieldbrewery.com.
https://jeremiahstaproom.co.uk.
www.crossborders.beer/old-eastway-tap.

 WHAT TO SEE & DO

Arrive at Waverley Station and the division of Scotland's capital into two main parts is immediately apparent. Turn north into the New Town or go south to the **Old Town**, which is where you should probably start, specifically at **Edinburgh Castle** on Castle Rock because it will give you a perspective on everything that comes after. The castle's structure as well as its exhibits tell stories about war – in general, for independence, and to see off Napoleon – and there is a Great Hall where James IV partied and Oliver Cromwell billeted his men. At 1pm most days a gun is fired, according to tradition, so that they can hear what the time is down on the Firth of Forth. It will also signal lunch. www.edinburghcastle.scot.

The Castle marks the starting point of a route along the ridge that passes **St Giles' Cathedral** [3] and the **Scottish Parliament** and ends at the **Palace of Holyroodhouse**, all must-sees. Brace yourself: the Royal Mile is tartan tat central but has some interesting stops along the way. **The Writers' Museum** [5], in one of Edinburgh's historic closes, reveals in part why this is a UNESCO City of Literature with its tribute to three Scottish literary giants, Robert Burns, Sir Walter Scott and Robert Louis Stevenson. Other writers get an inscription on a paving stone outside, while Harry Potter fans will be more excited by nearby **Victoria Street** [4], said to have inspired JK Rowling's Diagon Alley in the books.

www.stgilescathedral.org.uk.
www.parliament.scot.
www.rct.uk/visit/palace-of-holyroodhouse.
www.edinburghmuseums.org.uk/venue/
    writers-museum.

A short diversion south will take you to the **National Museum of Scotland**, a comprehensive appraisal of the nation's culture and natural history that's easy to digest and set in a lovely glass-topped building influenced by the Crystal Palace. Back on the Royal Mile, the stories continue at **The Real Mary King's Close**. It may be touristy but this old street discovered beneath the Royal Exchange is still intriguing (and a location in the novel *Mortal Causes* by Edinburgh author Ian Rankin). Further along, **The People's Story Museum** champions oral as well as written histories of the city's working-class community.

www.nms.ac.uk/national-museum-of-scotland.
www.realmarykingsclose.com.
www.scottishstorytellingcentre.com.
www.edinburghmuseums.org.uk/venue/
    peoples-story-museum.

WHILE YOU'RE HERE

Pull on your walking boots and climb Edinburgh's mountain, Arthur's Seat.

www.historicenvironment.scot/visit-a-place/
    places/holyrood-park.

[3]

[4]

[5]

No 1
WEST REGISTER STREET
GUILDFORD ARMS
THE
PUBLIC BAR
RESTAURANT
PUBLIC BAR
FAMOUS
AWARD WINNING
FREE HOUSE
SERVING
THE CITY'S
BEST CHOICE
OF SCOTTISH
REAL ALES
OWNED &
OPERATED
BY THE
STEWART FAMILY
SINCE 1898

NEW TOWN

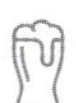 WHERE TO DRINK

A short walk from Edinburgh Waverley, off Princes Street, the **Guildford Arms** [6] has long been a favourite destination among beer lovers. Amid its Victorian opulence, 10 handpumps pour some of the best examples of Scottish cask ale, and there are tempting craft brews on the taps, too, these days including IPAs and authentic lagers. https://guildfordarms.com.

Pedestrianised Rose Street was traditionally Edinburgh's main drinking street and there's still plenty of choice here. The **Abbotsford**, at the Waverley end, with its panelled walls and ornate island bar, is probably the pick with some boldly modern cask choices on its half-dozen pumps. At the other end of the street, **Fierce Beer** is the local outpost of the eponymous Aberdeen brewer, with 20 lines pouring an adventurous array of craft beers. https://theabbotsford.com. https://fiercebeer.com/pages/edinburgh-bar.

Around the corner you'll find an Edinburgh institution. The **Oxford Bar** is an unspoilt backstreet pub, the main drinking area in a side room away from the bar itself. A changing roster of Scottish cask beers add to the appeal, while the 'Ox' has become famous in recent years as the local of crime writer Ian Rankin, who has his fictional detective Rebus drink there, too, for convenience. www.oxfordbar.co.uk.

Kay's Bar [7] is less well known but just as characterful, in its own way. Previously trading for 150 years as a liquor store in the roughest part of town, now barrels line the walls and the whole place is bathed in a red glow, appropriate to beer drinking. Six pumps feature mostly Scottish-brewed cask beers. www.kaysbar.uk.

On the Water of Leith path, **Stockbridge Tap** is a serious specialist beer house. The cask pumps favour Scottish brewers such as Swannay and Cromarty, while the keg lines roamed as far as London's Kernal and Worthing's Hand Brew Co at last visit. www.facebook.com/thestockbridgetap.

The word 'new' is rather misleading because this area was built in the late 18th and 19th centuries, the jumble of lanes around the castle having exhausted the space available to grow on the ridge. The first streets set out an orderly plan of wide avenues and garden squares and circuses named for royalty and the Union, with generously proportioned townhouses for the wealthy. As you wander, you'll find the **Scottish Gallery**, the national **Portrait Gallery**, and the National Trust for Scotland's **Georgian House**.
https://scottish-gallery.co.uk.
www.nationalgalleries.org.
www.nts.org.uk/visit/places/georgian-house.

At the western edge of this ordered society, the **Water of Leith** meanders to the sea. Pick up its path at Stockbridge to see one of sculptor Sir Antony Gormley's figures, **6 TIMES**, standing in the river here. Continue upstream beneath the high arches of a bridge designed by Thomas Telford, which amplifies the rush of the river to a roar, to arrive in **Dean Village**. This milling community, dating from the 12th century, is an extraordinary sight in a city centre, a strikingly peaceful scene of cobbled streets lined with old grain mills, cottages and warehouses that terrace up the hillsides above the leafy riverbanks. Continue onwards to the national **Modern** galleries or return to Princes Street, a few minutes' walk away.
www.waterofleith.org.uk.
www.nationalgalleries.org.

Where once there was a cesspit, now stands **The Mound**. Well, they had to dump the soil from the New Town somewhere, so they drained putrid Nor Loch below the castle and created a link to the Old Town. The Mound provided a base for a green corridor, **Princes Street Gardens**, the massive neo-Gothic shrine the **Scott Monument**, dedicated to local author Sir Walter Scott, and the **National** gallery, where Titians and Rembrandts hang alongside work by Scottish artists whose names might not trip off the tongue but are stars of a recently opened gallery showing the nation's artistic contribution from 1800 to 1945. There's more Scottish art down the side of Waverley station at the **City Art Centre** and international contemporary art at **Fruitmarket**.
www.edinburghmuseums.org.uk.
www.nationalgalleries.org.
www.edinburghmuseums.org.uk/venue/
    city-art-centre.
www.fruitmarket.co.uk.

WHILE YOU'RE HERE

Visit **Edinburgh's Royal Botanic Garden** [9], where the Victorian Palm Houses have been restored. Climb **Calton Hill** [10] to see the unfinished neoclassical National Monument of Scotland on its brow and fine views of the city and the Firth of Forth.
www.rbge.org.uk.
https://ewh.org.uk/calton-hill.

[9]

[10]

[11]

LEITH

WHERE TO DRINK

No fewer than four breweries occupy Edinburgh's docklands, each with on-site taprooms, and there are some great pubs, too, so it's well worth taking a tram up here. First stop, off Leith Walk, is the highly regarded **Newbarns Brewery** [11]. Its taproom, discreetly signposted at the end of an alleyway, is a relaxed haven that attracts a diverse cross-section of the community with a range of beers made on site including lagers, pales and stouts.
https://newbarnsbrewery.com.

Nearby, opening with more limited hours, are **Campervan Brewery**, with 10 beers on tap including guests, and **Pilot Brewery**, which is always experimenting with new recipes and offers through ScotBeer Tours an educational experience on Saturdays that includes a guide to local pubs.
www.campervanbrewery.com.
https://pilotbeer.co.uk.

Moonwake Beer Co. is found in an old warehouse further north by the Water of Leith. Its colourful taproom on a mezzanine above the brewhouse pours a dozen of its beers that might include a milk stout or a gose alongside pales and stouts. Brewery tours are offered, too.
www.moonwakebeer.com.

Carriers Quarters is a quirky little place with a bar in the front room and a corridor leading to a lovely space at the back decked in old advertising signs, while the 18th-century **Malt & Hops** is one of those warm pubs that draws you in with a real fire and a range of real ales to match. Between them, the **Innis & Gunn** taproom offers a rather different experience with 16 rotating craft beer lines pouring in a big modern bar.
www.carriersquarters.co.uk.
www.facebook.com/realaleleith.
www.innisandgunn.com/taprooms/edinburgh/leith.

Across the bridge, within the quayside development, **Campervan Brewery** operates **Lost In Leith**, a spacious cellar-style bar with 18 taps pouring a selection of its own brews and showcasing a wide variety of beers from across the world. You'll find barrel-aging going on here, too, in three giant oak foeders, and Campervan's own Beer Academy offering tutored tastings of different styles. Around the corner **Teuchters Landing** [12] enjoys a perfect spot on the water's edge with plenty of outdoor seating where you can enjoy a local cask ale.
www.campervanbrewery.com/lostinleith.
https://teuchtersbar.co.uk.

Before leaving Leith, you must take a walk up to the **Dreadnought**, an apparently ordinary pub on the corner of terraced streets where you'll open the doors on a truly welcoming local with some weird beer fonts pouring some well-chosen brews from cask and keg, usually including two or three from Brass Castle Brewery in Durham – owned by the brother of the landlord here.
www.dreadnoughtpub.com.

[12]

WHAT TO SEE & DO

Once, few visitors ventured to Leith except devotees of *Trainspotting* keen to see the underbelly of Edinburgh as described by Irvine Welsh and filmed by Danny Boyle. Now this part of town is firmly on the tourist map, regeneration having advanced at pace, creating a new profile for the old port as a place to find 21st-century café society. The new tram link brings a steady flow of people from the city centre for shopping, drinking and dining (to Michelin-star standard), pop-up events at **The Biscuit Factory** and **Leith Theatre**, exhibitions at the **Edinburgh Sculpture Workshop**, and tastings at the UK's first vertical whisky distillery, the **Port of Leith Distillery**. If it all gets a bit much, you could book a guided tour with **Invisible Cities** and strip off the veneer. Otherwise, Leith is light on attractions other than the biggie, to go aboard the **Royal Yacht** Britannia and see the style in which royalty ride the ocean wave.
www.biscuitfactory.co.uk.
www.leiththeatre.co.uk.
https://edinburghsculpture.org.
www.leithdistillery.com.
https://invisible-cities.org.
www.royalyachtbritannia.co.uk.

FURTHER INFORMATION

For more ideas on what to see and do, go to https://edinburgh.org and www.visitscotland.com. Check opening times of beer venues and attractions before you go.

Edinburgh Festival: The world's biggest celebration of the arts takes over the city every August with superstars and student theatre troupes competing for attention among the thousands of performances that take place on stage and street.
www.eif.co.uk, www.edfringe.com.

GLASGOW

Glasgow has a proud brewing heritage, and some famous pubs alongside some great modern bars. It has some great sights to share, too, but this working city comes to life when it tells the story of its people.

 WHERE TO DRINK

Tennent's Lager certainly has its place in brewing history having been brewed on the same spot in Glasgow's Wellpark since 1885. Today the city celebrates its heritage as home to one of the first British-brewed lagers with a tour and a beer masterclass at the brewery – and fitting that it's within sight of a modern manifestation of beer culture.

Drygate Brewery launched as a joint venture between **Tennent Caledonian Breweries** and Alloa's **Williams Bros Brewing** under the saw-toothed roof of a former box factory in 2014. It's a big buzzing brewpub and brasserie across two floors. Around 20 taps pour beers brewed here and at selected craft brewers around the UK. Tours and tutored tastings are available on Sundays. www.drygate.com. www.tennents.co.uk/experience/masterclasses.

In recent years, Glasgow has also become a centre for more experimental brewing. To the north of the city, alongside the canal, **Epochal Barrel Fermented Ales** is an exciting project drawing on Scotland's rich brewing history and modern hops to create some extraordinarily complex beers. On Fridays and Saturdays since July 2024 it has been possible to taste them at the onsite taproom where you can also gaze lovingly at a Burton Union brewing set rescued from Marston's. epochal-barrel-fermented-alesmyshopify.com.

Perhaps the city centre's most famous watering hole is the **Horseshoe Bar** [13]. Best known locally for its (almost) nonstop karaoke sessions upstairs, downstairs is relatively peaceful with, allegedly, the longest bar in Europe and some decent cask ales on rotation. Around the corner, the **Drum and Monkey**, part of the Nicholson's chain, which specialises in historic interiors, is a reliable spot for a pint of cask and also serves Innis & Gunn's lager from the tank, while the **Pot Still** is a warmly welcoming traditional pub that just happens to have 800 different whiskies lining the backbar. That doesn't distract them from serving a good pint of locally brewed cask ale, mind, and some great pies. www.innisandgunn.com/taprooms/glasgow/ city-centre. www.nicholsonspubs.co.uk. https://thepotstill.co.uk.

[13]

Glasgow's big surprise is the **Shilling Brewing Co** brewpub housed in the grand marbled walls of a former bank. It claims to offer the widest selection of draught beers in Scotland with more than 30 lines pouring, including its own brews, which ferment in the vaults beneath your feet. Tours and tastings are available, and there are pizzas for the peckish. **Innis & Gunn** also has a modern taproom nearby, rivalling the Shilling range with its 21 taps, including some of it's own limited editions and unpasteurised lager fresh from the tank. It runs Brew Schools, too, where you can come away with a 5-litre mini-keg of your own beer.
https://shillingbrewingcompany.co.uk.
www.innisandgunn.com/taprooms/glasgow/
    city-centre.

To the west of the city centre, in the Anderston area, there is a cluster of pubs worth visiting. **Bon Accord** is a traditional family-run ale house that serves 800 different cask beers a year from across the UK. **The Griffin** is the Isle of Skye Brewery's tap in the city, where you can taste its cask and keg beers. And in **State Bar** you can find up to six pumps pouring interesting cask ales from craft brewers in glamorous Victorian surroundings.
www.bonaccordpub.com.
https://thegriffinglasgow.co.uk.
State Bar, 148 Holland Street, G2 4NG.

Before you leave Glasgow, you must take time to go to the **Lauriston Bar**, just across the Clyde from the city centre. It's an almost unspoilt 1960s pub, attracting film crews, and has assumed a cult status with its red Formica-topped tables surrounding an island bar, its jukebox and its Scotch pies. The best thing is that its beer range is very much up to date, including brews from the likes of Fyne Ales. Southside is also home to **Koelschip Yard**, a craft bar focusing on beers from independent brewers around the world, with 14 lines of keg and one cask.
www.facebook.com/TheLaurieston.
www.koelschipyard.beer.

Gareth Young, Epochal

Gareth Young is a genuine philosopher-brewer. He might have given up his job at the University of Glasgow to brew full time, but he's creating beers with big ideas behind them. After winning the National Home Brewing Awards while a PhD student, he founded Epochal Barrel Fermented Beers in 2021, delving into Glasgow's brewing history to rediscover old techniques for making modern beers in oak.

'It's an engineering tradition rather than a farming tradition,' he says. 'We use some unusual processes here, encouraging aged hops to interact with the fermentation to generate highly complex ales, for instance, and making very high gravity beers using a special type of mashing vessel. We have started brewing more sessionable draught beers at 4 to 5% abv for pubs, too, though.'

Gareth believes there are 'more exciting things happening now in Glasgow than ever' when it comes to brewing innovation. 'New small brewers such as Simple Things Fermentations and Dookit Brewing have a strong sense of their local identity. Along with a certain style of traditional pub with the horseshoe-shaped bars, it makes Glasgow a special place for beer lovers.'

https://epochal-barrel-fermented-ales.
myshopify.com/pages/about.

👓 WHAT TO SEE & DO

Understanding the past explains what we see today, and perhaps no more so than on the Clyde in Glasgow, where the **Finnieston Crane**, 'the cran', now stands as a silent reminder of why this city grew up around the banks of the river. It was once the centre of activity, fitting engines in the ships built here and loading cargo onto vessels including locomotives from the railway works in the suburb of Springburn. The crane makes you wonder at how times have changed. Will the neighbouring events arenas, the **Armadillo** and **Hydro** by Norman Foster, the kind of structures you see in any city with the means to commission them, be regarded as emblematic of this city by future generations?
https://bigcranco.co.uk.
www.sec.co.uk.
www.ovohydro.com.

The best place to begin to understand the importance of the crane and the Clyde is along the river at the **People's Palace** [14] on Glasgow Green. This museum is the template for how to tell the story of a place, with its joyful focus on the people who built the city. Glasgow's development dovetails with tales of struggles such as Red Clydeside, when a wave of industrial militancy and political radicalism swept the docks in the early 20th century. The everyday things that have formed Glasgow's character are revealed, too. It must raise smiles and shudders from the locals to see reminders of the tenements that housed the booming population in Victorian and Edwardian days, the 'steamies' where the weekly wash was done, and dancing to The Gaybirds at Barrowland Ballroom.
www.glasgowlife.org.uk/museums/venues/
peoples-palace.

Barrowland Ballroom [15] is a short walk from the People's Palace. It was built by Maggie McIver to host entertainment for the stallholders from **The Barras**, the market she originally created as a gathering of handcarts in 1920 – barras as in barrows – and later replaced with stalls and a roof. Barrowland still hosts gigs and The Barras still does good business, probably because it's gone the way of most markets, becoming the domain of designers, makers and street-food operators and turning old buildings into creative spaces. https://barrowland-ballroom.co.uk. www.barrasmarket.com.

Head into the city centre and you'll meet Billy Connolly, or at least **a mural of him**, in Osborne Street (hopefully you saw his original banana boots back in the People's Palace). You'll also see some of many buildings created by **Alexander 'Greek' Thomson**, who earnt his nickname for the prolific Classical Grecian references in his designs, such as the Grecian Chambers on Sauchiehall Street. But Charles Rennie Mackintosh is Glasgow's favourite son. You can follow the progress of the Art Nouveau architect, designer and artist around the city on a self-guided trail, which includes his only church, **Mackintosh Queen's Cross**, and a stop for refreshment at **The Willow Tea Rooms** on Buchanan Street, which he designed for the temperance campaigner and entrepreneur Kate Cranston.

www.citycentremuraltrail.co.uk.
www.alexanderthomsonsociety.org.uk.
https://mackintoshchurch.com.
www.willowtearooms.co.uk.

The recreated interiors of Rennie Mackintosh's home, offer a fascinating insight into the artist's life at The Mackintosh House, in **The Hunterian**, and the biggest display of objects and furniture he created is on show in the **Mackintosh and the Glasgow Style Gallery** at the **Kelvingrove Art Gallery and Museum**. These institutions in themselves should be on your must-see list, along with **The Burrell Collection**, **Gallery of Modern Art**, **Glasgow Botanical Gardens**... How long are you staying?
www.gla.ac.uk/hunterian.
www.crmsociety.com/venue/kelvingrove-art-gallery-and-museum/.
www.glasgowlife.org.uk/museums/venues/kelvingrove-art-gallery-and-museum.
www.crmsociety.com.
https://burrellcollection.com.
https://galleryofmodernart.blog.
https://glasgowbotanicgardens.com.

FURTHER INFORMATION

For more ideas on what to see and do, go to www.glasgowlife.org.uk and www.visitscotland.com. Check opening times of beer venues and attractions before you go.

BABBITY BOWSTER

Down a quiet turning in the heart of Glasgow's Merchant City, Babbity Bowster opened in 1985 in an 18th-century town house that's attributed to Robert and James Adam – its name refers to an old Scottish country dance. There are six en-suite rooms upstairs, doubles and twins, recently refurbished in a bright modern style with textured finishes and walls depicting street maps of the city. Service is friendly and informal with breakfasts served in the pub downstairs. Babbity's hosts regular live folk sessions and has three cask ales on the bar plus a menu mixing traditional Scottish classics with international dishes. https://babbitybowster.com.

MORE PUBS WITH ROOMS

EDINBURGH

The Inn On The Mile [16]
www.theinnonthemile.co.uk.

Brewdog Doghouse [17]
www.brewdog.com/uk/doghouse-edinburgh-
   hotel.

Black Ivy [18]
https://weareblackivy.com.

GLASGOW

Rab Ha's [19]
https://rabhas.co.uk.

WALES

Opposite: Tenby harbour.

NORTH WALES COAST

The coast around Llandudno is one of the best stretches for beer in the principality, with its attractive pubs and vibrant taprooms, and while time has taken its toll on the seaside experience traditionally offered here, you'll discover an impressive castle, genteel streets and miles of golden sands.

CONWY

 ### WHERE TO DRINK

There are a few decent pubs in Conwy, but two stand out for beer. Near the train station, the **Bank of Conwy** [1] is, as the name suggests, a converted bank, retaining some original architectural features. It's got a surprising range of craft on a dozen taps including, at last visit, a couple from Llandudno's own Wild Horse, Cloudwater, Vault City and Beak, plus Surrey's Middle Child, Belfast's Bullhouse and Dutch brewer Uiltje. A couple on cask, too. Nearby, in a former ice-cream factory, is the latest and largest in the Tapps chain of pubs with a couple of dozen lines including eight cask.
Tapps, 6a Lancaster Square LL32 8HT
www.thebankofconwy.wales.

A little way up the road is a well-preserved example of a 1920s pub, the **Albion Ale House** [2]. As well as the main bar, a hatch serves a snug at the back and there's an open fire in the elegant lounge. The business is, unusually, a joint venture between four local brewers: Purple Moose, Great Orme, Bragdy Nant and Conwy Brewery. So expect to find their beers along with guests on 10 handpulls.
https://albionalehouse.weebly.com.

WHAT TO SEE & DO

If you were to design a harbour village, the ideal might look something like Conwy. Put the ruins of a **medieval castle**, all towers, battlements and curtain walls, at the water's edge with some misty mountains for a backdrop. Draw the outline of a circuit of **defensive walls**, with a few arches and some more towers, and within its boundaries sketch a grid of cobbled streets, populate them with rustic cottages and boutiques, and drop in an interesting building or two from down the centuries. For a finishing touch, add the wow factor; get an eminent engineer – Thomas Telford, say – to create a dramatic approach over a **suspension bridge** [4].

This little town on the west bank of the **River Conwy** [3], where Edward I threw up a formidable fortress in four short years during the late 13th century, is just such an idyll. It's hard to take your eyes off the castle and its walls and the way these structures frame harbour scenes. In fact, you could just focus on the fort and walk along the ramparts, which takes about half an hour depending on how long you are delayed by the views. https://cadw.gov.wales/visit/places-to-visit/ conwy-castle.

But there's more to see at ground level, and not just the cottage on the quay that holds the record for being the **Smallest House in Great Britain**, or the National Trust's 14th-century **Aberconwy House**. Conwy's architectural boast is the Elizabethan townhouse **Plas Mawr** because it's so rare in its authenticity. It was saved for the nation in the 1990s and restored and reopened as a heritage centre. And don't miss the **Royal Cambrian Academy of Art**, tucked up a backstreet, which promotes Welsh art in changing exhibitions. www.thesmallesthouse.co.uk. www.nationaltrust.org.uk/visit/wales/ aberconwy-house. www.rcaconwy.org.

LLANDUDNO

 ## WHERE TO DRINK

Since 2024, Llandudno's player on the craft beer scene, Wild Horse Brewing, has been able to showcase its excellent beers at its own taproom next to the brewery a five minute walk from the station. There are 10 lines to choose from while a roster of street vendors provide the food.

The beer scene in this corner of Wales is also invigorated by a four-strong micropub chain. The original **Tapps** [5] is in the centre of Llandudno in a former cake shop. A busy beer board tells you there are up to 10 lines pouring craft, including a few from Wild Horse, and another five cask. On the east side of town, **The Ascot Tapproom** is part of the same group. It's a cosy sort of pub without a bar counter, suitable for a single conversation among drinkers choosing from four lines of craft and two of cask.
www.facebook.com/TAPPS35.
www.facebook.com/Theascottapproom.

It's a real surprise to find a pub like the **Cottage Loaf** behind the town's giant Wetherspoon's. With its flower-decked terrace you could be out in the country if it wasn't for the view of the town hall car park. Inside the stone walls of this former bakery is a bustling food pub with an open fire and a bar serving five cask ales, including Conwy's Welsh Pride, plus craft cans from other Welsh brewers. https://the-cottageloaf.co.uk.

In the backstreets at the bottom of Great Orme, the **Snowdon** pre-dates Llandudno as a holiday resort but feels quite modern. It has five cask beer lines, usually including Draught Bass, and Wild Horse's pilsner-style lager Buckskin on the taps. thesnowdon.co.uk.

WHAT TO SEE & DO

Do you see this town as a magnet for OAP coach tours or a well-preserved example of a Victorian seaside resort? It's both, of course, but too easily dismissed as the former. Many seaside towns could learn about restraint from Llandudno because it has jealously guarded its sweep of 19th-century villas and hotels behind the

promenade. They still command the best sea views, unencumbered by arcades and tourist shops. Even **Venue Cymru**, a theatre, conference centre and arena big enough to hold several thousand people, manages to blend in. Seaside tat is confined to the pier and the streets behind the front (along with some good stuff, such as the **Mostyn** contemporary art gallery).
www.venuecymru.co.uk.
https://mostyn.org.

Llandudno's location on a headland meant it swerved the Expressway, the motorway-style stretch of the A55, which both serves and spoils this coast. That headland, the **Great Orme**, also provides the best reason to come here. This lump of limestone and dolomite has been a centre of tourism since the first season in the 1800s. Today a **cable car** and **tramway** [6] haul visitors up and down its mass, while those fit enough can walk to the top. On its slopes are a **stone circle** (recently installed), **botanical gardens**, **artificial ski slope**, **copper mines** turned tourist attraction, **nature reserve**, and **visitor centre** – which includes a room dedicated to UFO sightings because, apparently, there are so many. There's also the 12th-century **St**

Tudno's Church, which could feel remote if such a lot of former residents hadn't chosen to be buried around it. Oh, and a flock of Kashmir goats, which sometimes foray into town to raid the locals' privet hedges. Inevitably, the best attraction on the Great Orme is the view from the top.
Great Orme Cable Car, Happy Valley, LL30 2LP.
https://greatormetramway.co.uk/en/great-orme.
www.greatormemines.info.
www.jnlllandudno.co.uk.
www.nationalchurchestrust.org/church/
  st-tudno-great-orme.

The scampish Playing Cards from *Alice's Adventures in Wonderland* dance around the bandstand railings on the lawn below the botanical gardens. Down in the town, a statue of the White Rabbit looks ready to hop off his plinth in the North Western Gardens. They're part of an **Alice in Wonderland Town Trail** set out in 2015 to commemorate the 150th anniversary of the book by Lewis Carroll, aka Charles Dodgson. Dodgson is said to have met up with young Alice and her family, the Liddells, on the **West Shore** while on holiday here.
www.visitconwy.org.uk/things-to-do/alice-in-
  wonderland-town-trails-p293491.

[6]

RHOS & COLWYN BAY

WHERE TO DRINK

Tapps Rhos, the third micropub in the Tapps chain, is just off the coast road near Rhos on Sea Harbour. There are four cask lines here, featuring local brews including Conwy, supplemented by six craft keg led by Wild Horse. Look out for the Cheese Nights! www.facebook.com/TappsatRhos.

In Colwyn Bay, close to a Wetherspoon's in the old cinema, the **Bay Hop** is another micropub, this time with less emphasis on local beers. Ales on the four handpumps might come from anywhere in the UK, while the seven keg lines feature Belgians alongside well-known UK craft names plus, at last visit, a pilsner from Welsh German-style brewer Geipel. https://www.facebook.com/thebayhop/.

Ink is a curious addition to the local drinking scene. It's an art gallery with a bar pouring a couple of Wild Horse beers alongside a fridge with a small but well-chosen range of craft. Enjoy them is a relaxed and colourful café surrounded by paintings. www.inkgallery.co.uk.

Black Cloak Brewery & Taproom is a traditional sort of brewpub except that its own beers, brewed on a tiny kit out the back, are generally keg IPAs that line up with the likes of Donzoko, Sureshot and Rivington on a total of a dozen taps. Three cask options tend to be modern styles from craft brewers, too, and there's a good range of alcohol-frees in the fridge, mostly Mash Gang. www.facebook.com/blackcloakbrewing.

Two more places nearby worth a mention are the **Crafty Fox**, a micropub in a former shop in Old Colwyn which has earned a good reputation for both cask and craft and is bringing some interesting beers to the village, and, completely different, **Pen-y-Bryn**, a Brunning & Price gastropub where you can expect to find some decent cask ales as well as panoramic views of the coast and a lovely beer garden. www.facebook.com/TheCraftyFoxPub. www.brunningandprice.co.uk/penybryn.

Finally, tricky to find (so check the map on the website) off Tay-y-Graig Road in the countryside to the east, the **Conwy Brewery** taproom acts as a lively local for the people of Llysfaen and offers brewery tours to visitors at the weekend. www.conwybrewery.co.uk.

WHAT TO SEE & DO

The coast road out of Llandudno rounds the **Little Orme**. You will have to walk up this one, but you'll be spurred on by the promise of views, kittiwakes on the wing, and grey seals playing in waves below. The road continues to **Rhos on Sea**, where the lucky residents of the neat houses and bungalows that line it look out on wide sands. A rough stone chapel, one of the UK's smallest 'churches', stands by the shore, dedicated to 6th-century **St Trillo** [8]. The tiny grotto-like structure has been built and rebuilt since the 13th century and today's custodians keep it as a contemplative place, candlelit, with rustic seats before the little altar.

www.llandudno.com/little-orme/.
www.nationalchurchestrust.org.

Generous sands continue to **Colwyn Bay** [7], but the houses give way to wooded banks. The promenade and beach have undergone a multimillion-pound facelift with the addition of a small ornate section of **pier** salvaged from its derelict predecessor, and the massive modern edifice of **Porth Eirias**, home to a watersports shop and a restaurant by celebrity chef Bryn Williams. Is this a sign that Colwyn Bay is on the up? On the high street, historic **Theatr Colwyn** stages plays and puts on films. It also contains the Oriel Gallery, dedicated to photography. www.pierbaecolwyn.org. https://portheirias.com. https://theatrcolwyn.co.uk.

FURTHER INFORMATION

For more ideas on what to see and do, go to www.visitconwy.org.uk and www.visitwales.com. Check opening times at beer venues and attractions before you go.

[8]

PEMBROKESHIRE

Pembrokeshire is a wilder affair, with rocky shores and quiet valleys, where pilgrims worship at an oceanfront cathedral and quaint fishing villages provide a pretty setting for a beach holiday. Tenby is a gem for beer with two modern breweries, and Cwm Gwaun to the north enjoys its own brewing scene and some unusual pubs.

TENBY

 ## WHERE TO DRINK

Tenby is a buzzing holiday resort with a characterful old harbour that has attracted a pair of craft breweries and some interesting venues to explore. In the centre of town, **Harbwr Brewery** occupies a whole complex of former warehouses. The brewery itself is at the hub on a cobbled alleyway, and you can order a cask or keg beer from a hatch and take it to your choice of tables in the mazey courtyard opposite, where there's also a well-stocked shop. Or climb the stairs for the **Tap & Kitchen** and a sun-trap roof terrace. Or find your way back down again to the **Hope & Anchor**, a traditional pub that looks like a separate business from the outside. Multiple venues in one, and each very different. https://harbwr.wales.

Tenby Brewing Co., which leans more towards modern craft styles, has its headquarters on the edge of town where the expansive **Tap Yard** pours the full range during the summer months. It has also teamed up with Feast Pembrokeshire to create a bar-restaurant called **Tap & Tân** in The Mews arcade. As well as a wide choice of interesting brews from the taps, you can sit down to some imaginative fire-cooked food. www.tenbybrewingco.com. www.tapandtantenby.co.uk.

Both breweries have a presence, too, in Tenby's hinterland. Along the coast at Saundersfoot you'll find **Harbwr Bar & Kitchen**, a modern space with a terrace looking out to the harbour, while Tenby's **Hwb Tap & Food Hall** is inland at Narberth, offering 18 lines of craft beer to wash down street food cooked by various local partners. harbwrbarsaundersfoot.wales. www.hwbnarberth.com.

WHAT TO SEE & DO

You know you're on holiday when you arrive in a seaside town like **Tenby**, and not just because of the crowds. Clinging to a headland, fringed with golden sands, this town's pastel-painted Regency

terraces peep over the harbour wall at boats bobbing in the teal waters, the remnants of a medieval fortress still keeping an eye out for invaders. It all promises a jolly time.

And it delivers. When you're not busy with your bucket and spade on one of the three beaches, you can walk across the sands to **St Catherine's Island** at low tide for, well, a good look back at Tenby, or jump on the boat to see the monastery on **Caldey Island** [9]. Within the medieval town walls, along lanes crowded with fishermen's cottages and more Georgian townhouses that confirm you're not the first to happen upon Tenby's charms, there is the wagon ceiling to crane your neck at in **St Mary's Church** and Tudor Tenby to learn about at **Tudor Merchant's House** (Henry Tudor is said to have fled here during the War of the Roses).
www.caldeyislandwales.com.

www.stmaryschurchtenby.co.uk.
www.nationaltrust.org.uk/visit/wales/tudor-merchants-house.

A statue of Prince Albert suggests the Victorians also liked it here – so much, in fact, that they opened **Tenby Museum & Art Gallery**, which shows paintings by Gwen as well as Augustus John (they grew up here). However, you can't see the Victorian pier any more, it has been replaced by a huge **RNLI station**, which offers moments of real drama when the *Haydn Miller* launches from the slipway. (Next to it is the old lifeboat station, now a private home – its owners' quest to repel the sea is recorded in an episode of Channel 4's *Grand Designs*.) Still want more action? This is coasteering country, so go grab a wetsuit at **Tenby Adventure**.
https://tenbymuseum.org.uk.
https://rnli.org.
https://tenbyadventure.co.uk.

CWM GWAUN

🍺 WHERE TO DRINK

The Gwaun Valley is the hub of good beer in this part of Wales. There are two breweries. The eponymous **Gwaun Valley Brewery** and its taproom occupies an old granary where it pours cask ales in traditional styles and hosts acoustic music sessions. It has a campsite, too. Also on a farm, **Bluestone Brewing Co** is more about modern craft, serving a range of mostly pales at its taproom and yard.
www.gwaunvalleybrewery.com.
www.bluestonebrewing.co.uk.

The highlight of any beer lover's trip to Cwm Gwaun, though, has to be the **Dyffryn Arms** – despite the fact it serves only one draught ale. Better known as Bessie's, it's somebody's front room rather than a pub, a throwback to the beerhouses of the 19th century. Service is through a hatch and the beer, always Bass, is poured straight from the cask into a jug before being decanted into your glass. Sadly, Bessie Davies died aged 94 at the end of 2023, having run the place for 70 years. Her family are now keeping her very special legacy alive.
Dyffryn Arms, Pontfaen Road SA65 9SE.

Up the valley, **Tafarn Sinc** was built in 1876 out of corrugated iron and is now owned by the local community, serving local ales and good food. The 'tin pub', as it's affectionately known around here, is decorated with farming implements, including a nasty pair of castration pincers, hams hanging from the ceiling and sawdust on the floor.
www.tafarnsinc.cymru.

WHILE YOU'RE HERE

Head north to Nevern and the **Trewern Arms**. The rambling 16th-century inn is one of those pubs doing a bit of everything – fine food, hotel accommodation, and a hub for walkers on the coastal path and those messing about on the river on which it stands. The local ales on the pumps are decent, too.
https://trewernarms.com.

🔭 WHAT TO SEE & DO

You can leave the crowds well behind in this quiet rural valley in the northern reaches of the Pembrokeshire Coast National Park [11], where the **River Gwaun** makes its way from the **Preseli Mountains** to the sea at Fishguard. If you're in the village of **Pontfaen** on 13 January you'll witness the locals celebrating New Year – they keep the Julian calendar here – and children going from door to door, singing a traditional New Year song. Little changes here, which will make you slow down, too, and find it's enough just to follow the 9.7km (6-mile) trail on foot or bike through the steep-sided valley.
www.pembrokeshire.gov.uk/cycle-
    pembrokeshire/cycle-pembrokeshire-gwaun-
    valley-trail.

WHILE YOU'RE HERE

Not your usual hump in the ground, the stones of the 5,000 year-old Neolithic burial chamber Pentre Ifan [10] are laid bare to the elements.
https://cadw.gov.wales/visit/places-to-visit/
    pentre-ifan-burial-chamber.

ST DAVIDS

 WHERE TO DRINK

Bluestone Brewing has a joint venture in the smallest city with Newport's The Canteen. **Grain** has keg lines pouring Bluestone beers plus guests, while The Canteen supplies the pizzas in a rustic setting. Best traditional pub for beer is the **Farmers Arms**, a cosy nook pouring well-kept Welsh ales.
www.facebook.com/grainstdavids.
http://farmersstdavids.co.uk.

A little way up the A487 the **Old Farmhouse Brewery** grows its own barley in the surrounding fields and brews with water from its own well to make a range of bottled beers in both traditional and modern styles. It welcomes visitors and hosts occasional events.
www.oldfarmhousebrewery.co.uk.

WHAT TO SEE & DO

The crowds gather at Britain's smallest city, St Davids, to see the huge **cathedral** [12] dedicated to the saint who built a monastery on this spot in the 6th century. Set just below the village-city, as you descend the 39 steps from the Tower Gate you might think the cathedral slid down to its lower position at some point due to its great weight. The first cathedral was built in the 12th century; today's version is a bit later and has been added to, not least by Sir George Gilbert Scott. There's much to see inside, from the oak ceiling hung with pendants to a carving beneath a misericord of pilgrims being sick over the side of a boat, as well as the **Bishop's Palace** [13] on the opposite bank of the River Alun.
www.stdavidscathedral.org.uk.
https://cadw.gov.wales.

While the cathedral is the main attraction, the **Oriel y Parc Gallery and Visitor Centre** also deserves your time. It's an interesting fusion; a place to learn about the Pembrokeshire Coast National Park, to see pieces from Amgueddfa Cymru – Museum Wales, and to view a changing programme of art on the theme of local environment. Choughs are among the birds to see on a boat trip to **Ramsey Island**, as well as one of Britain's largest colonies of Atlantic Grey seals. www.rspb.org.uk/days-out/reserves/ramsey-island.

[13]

You might also find **Solva Woollen Mill's** shop [14] on the high street hard to resist, which sells beautiful textiles weaved nearby with Welsh tapestry designs. www.pembrokeshirecoast.wales/oriel-y-parc. www.solvawoollenmill.co.uk.

WHILE YOU'RE HERE

On the road to Haverfordwest, the **Victoria Inn Brewhouse** at Roch is an award-winning brewpub with an impressive range of beers on cask and in bottle from traditional ales to West Coast IPAs and tropical pales. There's a shop on site for takeaways and B&B accommodation if you really can't leave. www.thevictoriainnroch.com.

FURTHER INFORMATION

For more ideas on what to see and do, go to www.visitpembrokeshire.com and www. visitwales.com. Check opening times at beer venues and attractions before you go.

[14]

Y CAPEL

Characterful Y Capel is 15m (49ft) from the Erskine Arms in the centre of Conwy and part of the same business. Its name translates as The Chapel, and the 19th-century building retains many original features of that place of worship, including roof timbers, stone windows and a baptismal pool you can see under the floor at the entrance. The 11 comfortable boutique rooms, 10 double and one twin, are decorated in light, soft tones and spread over two floors of the chapel and the adjoining stable. Full Welsh breakfasts are served in the Erskine Arms, which has another 10 guest rooms upstairs. There are usually six cask beers on the pumps featuring mostly local brews, and a varied menu of home-cooked dishes. https://ycapel.co.uk.

STACKPOLE INN

Stay at the Stackpole Inn and you'll not only be able to make trips into Tenby, St Davids and Cwm Gwaun, but be close to one of the best beaches in Pembrokeshire (and there are a few), Barafundle. The 18th-century inn cloaked in creepers offers four doubles, two of which have sofa beds if you're bringing the family. The bedrooms in the annexe let in lots of light and have a soft, fresh air, dressed in coastal colours, with white wood panelling on the walls, while little touches such as a cushion decorated with a puffin help evoke a seaside ambience. The Stackpole has a serious foodie rep and at the bar you can expect Felinfoel alongside other Welsh brews. www.stackpoleinn.co.uk.

MORE PUBS WITH ROOMS

CONWY

The Erskine Arms [15]
https://erskinearms.co.uk.

The Bridge Inn
https://bridgeinnconwy.co.uk.

The Castle Hotel
https://castlewales.co.uk.

LLANDUDNO

Gwesty Links [16]
www.linkshotellllandudno.co.uk.

TENBY

Billycan [17]
www.billycan-tenby.co.uk.

CWM GWAUN

Llys Meddyg Hotel [18]
Newport, www.llysmeddyg.com.

ST DAVIDS

Grove Hotel
www.grovehotelstdavidspembrokeshire.co.uk.

[15]

[16]

[17]

[18]

PHOTO CREDITS

The publisher would like to thank the following for their kind permission to reproduce their photographs:

11 Stuart Brookes. **12** Marketing Sheffield. **16** York Castle Museum. **17** Rachel Auty. **23** Leeds City Council. **24** Kirkstall Brewery (left), Visit Leeds (right). **28** Visit Bradford (top), Tom Marshall (bottom). **31** Visit Bradford. **38** Cutlery Works. **40** Guy Fawkes Inn. **45** The Barrel House. **49** Stagecoach. **54, 55** Marketing Manchester. **60** David Dixon. **63** Alamy: Ian Pilbeam. **64** Adobe Stock: Torval Mork. **65, 66** Liverpool BID Company. **69** Brewhouse & Kitchen. **70** Chester BID. **71** iStock: Stuart Robinson Photography. **72** The Inn Collection Group (top). **72** Brewdog. **79, 81, 83** Visit Peak District & Derbyshire, 2024. **80** Thornbridge, 2024. **84, 89, 90** (top) Tom Platinum Monkey. **87** Dean Atkins (left). **88** Neon Raptor. **90** iStockL Robert Mackin. **92** Alamy: 2ebill. **93** Dreamstime: Clive Stapleton. **95** Visit Derby. **102** Remarkable Media Group / Sandwell & Birmingham Mela. **103** West Midlands Growth Company. **106** Daniel Capper (left). **106** Bespoke Inns / Graham Townsend (right). **108** Crate Brewery. **111, 113** Visit Norwich. **112** Hannah Hutchins. **116** Alamy: Kemaro. **118** Alamy: Clive Tully. **119** iStock: acmanley. **121, 124, 125** thesuffolkcoast.co.uk. **122** Alamy: Homer Sykes (top). Kevin Scott (bottom). **123** Alamy: R Kawka. **127** Alamy: Chris Batson. **128** 40ft Brewery. **129** Alamy: Jenny Matthews. **131** Crate Brewery. **132** iStock: Amanda Lewis (top). **132**. iStock: joningall (bottom). **134** Adobe Stock: Kemi. **135** iStock: VictorHuang (top left), Miles Willis (top right), Alamy: Michael Heath. **139** Greene King. **140** Adobe Stock: IWei (left), Lance Bellers (right). **141** Adobe Stock: Matthew Bugg. **142** Emma Kindred, 2019. **143** Alamy: Terry Matthews (top, centre). The Rose and Crown (bottom). **144, 160 (right), 161** Visit Lewes: Nigel French. **148** Alamy: Nigel Stack News. **149** Ian Price. **151** Visit Brighton (bottom). **154** Loud Shirt Brewery (left). **155** Adobe Stock: SarahLouise (bottom). **157, 159** Harveys Brewery. **158** Lewes Beer Mile (bottom). **160** Alamy: Eleventh Hour Photography (bottom). **164** Visit Kent. **165** Shepherd Neame. **166** Adobe Stock: sarahdoow. **167** Adobe Stock: Philip Carr. **172** Adobe Stock: Michalis Palis. **174** The Ramsgate Brewery, 2019 (left), David Anstiss (right). **175** The Ramsgate Brewery, 2019 (top); iStock: Moonstone Images (bottom). **176** Adobe Stock: Christine Bird. **177** Adobe Stock: Mike Higginson. **178** Alamy: Richard Milnes (right). **179** Alamy: J Marshall – Tribaleye Images (top), Bax Walker (middle top); Shepheard Neame (middle bottom). **182, 188 (left)** Good Chemistry. **184** Alamy: Urban Pubs. **185** Destination Bristol. **190** Jim Cossey. **191** Morgane Bigault. **194, 195** Visit Bath. **198, 199** Visit Cheltenham: Mikal Ludlow. **202** Stroud Brewery. **205, 211 (left)** St Austell Brewery, 2024. **206** Visit Cornwall: Matt Jessop (right). **207** Robert Herron for Verdant Brewing Co. **207** iStock: chrisdorney. **210** Paul Abbitt. **211** Alamy: Mr Standfast (top), Visit Cornwall: Matt Jessop (middle), Rikard Österlund (bottom). **213** iStock: Nigel Jarvis (top), Alamy: eye35 stock (bottom). **214** Alamy: Essential Stock. **217** Stuart Brooks. **219, 223 (right), 225** VisitScotland / Kenny Lam. **220** Alamy: DMac. **221** Alamy: Kay Roxby. **222** Alamy: Nick Servian. **223** VisitScotland / Paul Tomkins (left). **228, 229** Glasgow Life. **230** Alamy: Douglas Carr. **230** Alamy: Oscar Elias (bottom). **232** Adobe Stock: Kelly. **236, 241** Conwy County Borough / Ioan Said. **237** Conwy County Borough / Roger Richards. **238** LFX Media. **239** Conwy County Borough **240** Cymru Wales. **243** iStock: owainglyndwr. **244** Adobe Stock: Fulcanelli. **245** Alamy: Aurora. **246** iStock: HildaWeges. **247** Cymru Wales (top), Alamy: Ian Bottle (bottom). **248** Y Capel (left). Owen Howells (right). **249** iStock: Shawn Williams (top).